The MIT Guide to Science and Engineering Communication

second edition

The MIT Guide to Science and Engineering Communication

second edition

James G. Paradis and Muriel L. Zimmerman

The MIT Press
Cambridge, Massachusetts
London, England

This book was set in Sabon on 3B2 by Asco Typesetters, Hong Kong, and printed in the United States of America.

Library of Congress Cataloging-in-Publication Data

Paradis, James G., 1942–
The MIT guide to science and engineering communication / James G. Paradis and Muriel L. Zimmerman.—2nd ed.
p. cm.
Includes bibliographical references and index.
ISBN 0-262-66127-6 (pbk. : alk. paper)
1. Communication in science. 2. Communication in engineering. 3. Technical writing. I. Zimmerman, Muriel L. II. Title.
Q223 .P33 2002
808′.0665—dc21 2001056221

10 9 8 7 6 5 4 3

Contents

Preface to the Second Edition

In the five years since the first edition of this book was published, the practices of science and engineering communication have been transformed by computer technology. The distinctions between memoranda and letters are now blurred, and most correspondence is transmitted electronically. Proposals are submitted on-line, prepared with templates downloaded from agency Web sites. Reports are distributed to clients through intranets, and their content includes video and sound as well as traditional tables and figures. Journal articles are increasingly written for full electronic transmission. Conference abstracts are submitted through the Web sites of professional societies, and oral presentations are supported by computer-based slide presentations and later uploaded to an organizational Web site, available for review to interested parties who were not present at the conference. Résumés and curricula vitae are routinely submitted through e-mail and posted on the Web.

Writers in science and technology "attend" network meetings, use the information resources of the Internet, and have personal as well as organizational home pages. They work in companies that have replaced multivolume manuals with information provided on CD-ROM or the Web, perhaps to field technicians who use handheld computers at remote sites. They have ongoing relations with readers, providing updates rather than waiting for formal requests, participating in electronic conversations about their work, revising documents when better information becomes available. Every chapter of this second edition of *The MIT Guide to Science and Engineering Communication* reflects these changes.

The materials in this book are drawn from our teaching of scientific and technical communication to two different audiences. As faculty members at the University of California, Santa Barbara, and at the

Massachusetts Institute of Technology, we teach communication to engineering and science majors. As trainers in seminars in industry and government, we instruct scientists and engineers in professional settings. The materials we use in this book will, we hope, bridge the gap between the university novice and the seasoned professional.

Our approach is to emphasize specific processes and forms that will help individual writers create documents. We recognize, however, that writing takes place in the social context of local groups and larger organizations. Most writing in science and engineering is collaborative. Coauthored documents are cycled through editing and review and then often issued with a corporate name as author. Collaborative writing influences nearly every phase of the process; finished documents represent the work of many people.

Throughout this guide, we make a special effort to provide realistic examples from actual documents and situations. Most of our examples have already been used in college classrooms and professional seminars. Our experience is the basis of our book.

Acknowledgments

We are grateful to the many teachers, colleagues, and clients who have taught us, read our manuscripts, furnished examples, and given us advice. We appreciate the insights and concrete suggestions given us by our students at the University of California, the University of Washington, and MIT over the past two decades. We appreciate the support and advice of MIT Press editor Larry Cohen and the skillful artwork prepared by designers Stephanie Simon and Jim McWethy.

Jim Paradis thanks Jim Souther, Mike White, Robert Rathbone, John Kirkman, Peter Hunt, Steve Gass, John Kirsch, Ed Barrett, Marie Redmond, Harold Hanham, Anthony French, Tom Pearsall, Charles Bazerman, Charles Sides, Jim Zappen, Les Perelman, Dave Custer, Dan Cousins, Chris Sawyer-Laucanno, Bob D'Angio, Anne Lavin, Kenneth Manning, Leon Trilling, Frank McClintock, Jay Lucker, Tom Weiss, and Mary Pensyl, John Fothergill Jr., Maya Jhangiani, and Doug Bresh.

Muriel Zimmerman thanks Hugh Marsh, Saul Carliner, Jack Falk, Kenneth Manning, Alex Nathanson, Ellen Strenski, George Hayhoe, Roger Grice, Rudy Joenk, Gene Hoffnagle, Bernadette Longo, Marj Davis, Ron Blicq, Lisa Moretto, Ed Clark, Bill Kehoe, Beth Moeller, Luke Maki, Kim Campbell, Nancy Coppola, Tom VanLoon, Terrance Malkinson, and Cheryl Reimhold.

We are also grateful to the many engineers and scientists at sites including The Applied Physics Laboratory (University of Washington), Brookhaven National Laboratory, the Department of Interior, Department of Energy, Exxon, and Mitre Corporation for teaching us about the roles communication plays in the work of professionals.

Part I

1

Writing and Work

■

Consider this situation. A research group carries on an informal discussion with colleagues and management. Through the discussion, the group develops an initial concept for a new coal atomization process. This concept is presented in an in-house proposal to local management and then as a detailed proposal to a government sponsor. The project is funded, and the ideas are worked out in greater detail. Text, figures, and tables are recorded in researchers' notebooks and computer files. Some of this material furnishes the computer graphics for Thursday afternoon in-house seminars. Later still, the same notes, data files, and figures are recorded and circulated as progress reports to the sponsor. Eventually—after still more informal discussion, progress reports, and meetings—aspects of the researchers' coal atomization process take shape as one or more formal reports, journal articles, process specifications, patent applications, and design standards. Along the way, the group will have generated a good number of administrative and technical correspondence, most of it in the form of electronic mail.

The most effective scientist or engineer is typically a skilled writer. Communication skills are so essential to sharing the results of science and engineering that writing often becomes a large part of any job. As engineers and scientists move up the organizational ladder, to supervisor and then to manager, they spend more and more time on communication tasks, reviewing and editing the writing of their subordinates as they assume responsibility for meeting group objectives and deadlines. Independent consultants spend still more time preparing documents for their various clients.

Engineering and scientific communication is a fluid activity. Writing extends and complements other forms of work. It helps to shape and share thought processes, research records, specifications, decisions—anything that can be represented in words, symbols, or graphics. Documents are records of the steps of decision making, design, reasoning, and research. Writing is the preeminent means of transferring information and knowledge in detail and accuracy.

The Social Context of Scientific Writing

Scientific writing is social in two senses. First, it is typically collaborative, the result of teamwork among peers and management. Second, the written document itself circulates in a community of specialists. An internal review process helps writers shape information into useful arguments that address their projected readers. Collaborators may be colleagues, supervisors, or outside readers. They may contribute to the finished product. They may provide comments and information. Or they may guide and evaluate the work.

The reviewing process, as shown in Figure 1.1, has different implications in different environments. Student writing, for example, is rarely true collaboration and has no audience beyond the instructor. This way of learning sometimes leads the novice to underrate the importance of writing in the professional world. Workplace writing, on the other hand,

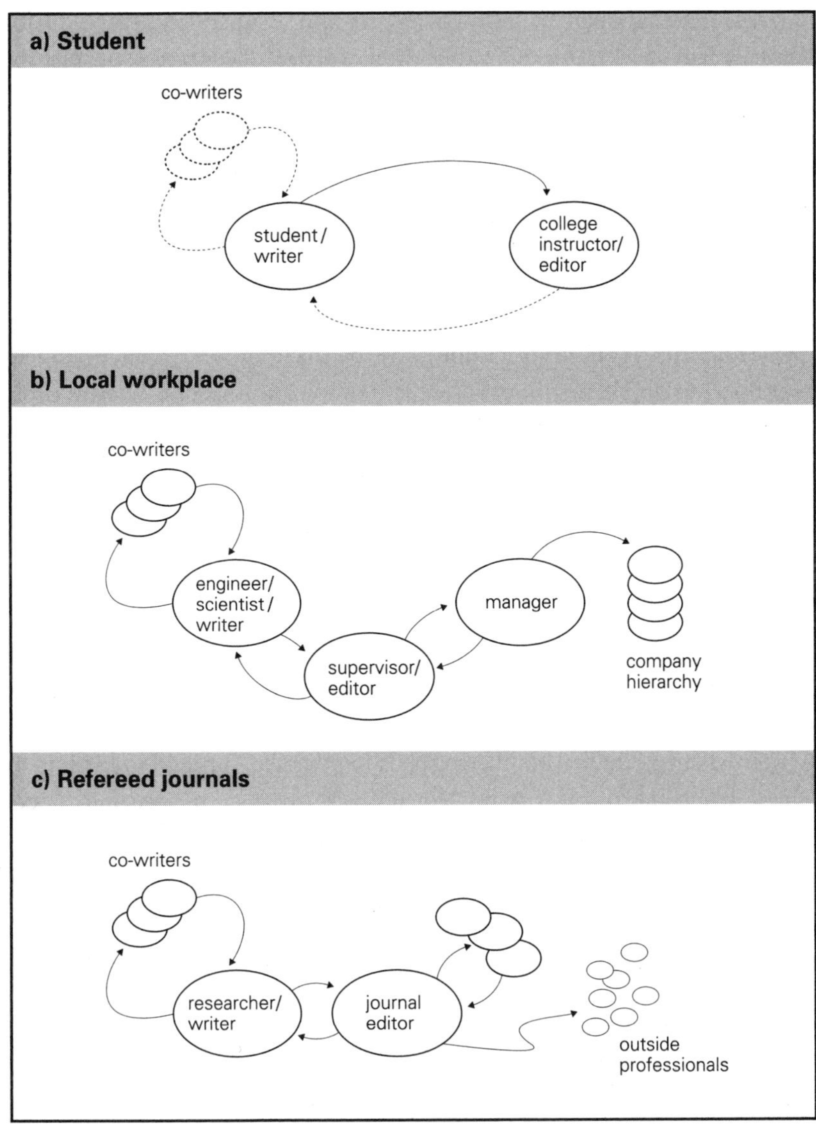

Figure 1.1
Three kinds of review cycles: (a) student to instructor, (b) employee to management, (c) writer to editor by way of expert referees. Student writing generally has no audience beyond the instructor; after supervisory review, workplace writing reaches company hierarchy; publications in refereed journals reach a wide audience of professionals in the same field.

is generally examined by both colleagues and a supervisor, who edit for content and style. In formal publication, the document passes outside the institution to a professional editor, who circulates it to referees and may ask for revisions.

The Politics of Written Communication

Most writing will have some political significance, quite apart from the primary message. To write is to assert, and assertions involve other people's interests. Your information may be accurate and your argument worthy, but you can still make big communication errors. Writers often do not appreciate the extent to which their activities impinge on the interests of others, whether in focusing a problem, developing a document plan, or drafting, revising, and producing a manuscript. At each of these stages, a writer needs to consult colleagues and supervisors—and perhaps to rework the initial efforts in order to develop an improved strategy for persuasion. The formal, permanent aspects of a written document may be inappropriate when a more personal touch is required or when a record isn't really needed. No matter what the technical merits of a written proposal, it may seem confrontational to management if the writer has neglected to build consensus in advance, through individual or small-group meetings. It is not always wise to rush an idea into document form; time can often be better spent discussing ideas and perhaps being prepared to share credit for innovations.

Recording as the Basis for Writing

It is sometimes tempting to think that comprehensive research precedes all writing. This is clearly not the case. Numerous writing and information-gathering activities take place while research is carried out, and these activities, in turn, furnish the basis of all project-related writing.

Consider a research project in which a physicist, physician, and medical technologist conduct a five-month series of experiments to study the pattern and extent of lithium distribution in sections of human brain. The investigators collect over 20 recent papers on lithium treatment of mania and depression, nuclear analytical procedures for analyzing lithium distribution, and modes of lithium action in rodent brain tissues.

They use a high-frequency beam reactor to bombard human brain tissue samples with neutrons, which cause a lithium isotope in the brain to release energetic particles.

They fill several notebooks with the details of the experimental design, methods of preparing cross sections of brain tissue, inscription records of the cross sections, data from particle detection, data reduction and rough graphs, notes on error analysis and sensitivity ranges for the experiment, and case histories of deceased patients who had undergone lithium treatment. Funded by a national health foundation, they are expected to prepare a report and to publish two or three papers on their findings in refereed journals read by clinicians and health researchers.

Like most research projects, this one generates an immense—and potentially chaotic—volume of written and visual detail long before any formal write-up of results takes place. The detail is a combination of previously published papers, a proposal, correspondence, photographs, spreadsheets, graphs, patient records, notebooks, and notes from meetings and informal discussions. This thicket of information needs to be sorted and arranged so that its patterns can be studied and it can be retrieved when necessary.

Effective writing requires initial organization, a task that writers sometimes underestimate. When information becomes available, you need to preserve it. The articles or reports you fail to file, the comments you do not record, the meeting notes you lose, the data you don't get around to entering, the files you fail to organize in the computer, the procedure you forget to write down—any of these lost or neglected items can haunt the researcher-turned-writer. Even small items—a missing reference, a physical constant, a procedural description—can turn a routine writing task into a guessing game. The failure to organize information as it's gathered accounts for many of the problems writers experience.

Planning a Recording Program

A program of information gathering, recording, and archiving is a way of anticipating the written and oral presentations that will inevitably follow. The ability to get to the various sources of information is essential to solving problems. Your design for arranging and storing material will save hours later and may well save you from having to reconstruct events from an incomplete or vague record. Here are some suggestions:

- Design a system that will arrange computer files for anticipated use in writing.
- Arrange published materials, correspondence, and other collectibles in file folders, loose-leaf folders, and vertical files.
- Keep a record of all meeting notes and agendas for future reference.
- Record experimental procedures, details, notes, and procedures in routinely updated laboratory notebooks.
- Sketch and arrange preliminary graphics in laboratory notebooks and computer files.

Using Notebooks

Although organizing records of your accumulating work may at first seem like drudgery, your records and files do assume great value with time. They are your personal store of information, extensions of your memory (Figure 1.2). Records require you to sort information conceptually. What is included and what is left out are matters of great significance.

Systematically kept, your notebook preserves the content and sequence of your activities. Your notebook makes it possible to reconstruct project developments. Always date the pages. A research record in a permanently bound notebook with printed page numbers is also a legal record of ideas, drawings, or descriptions. Maintain vertical files for material that does not fit in the notebook. Drawings, photographs, blueprints, equipment specifications, computer printouts, and calculations are all worth saving.

Items commonly recorded in notebooks include:

- Objectives: the purpose of an experiment and the time of day of the experimental activity
- Procedures: rough descriptions, sketches of apparatus, modifications to apparatus, steps in the procedure, notes on equipment and materials used
- Results: columns of data, rough graphs, descriptions, observations, photographs, printouts
- Analyses: equations, narrative comments, unanswered questions, data reduction techniques, new ideas, references to the published literature, correlations of data

Project record keeping is crucial. Laboratory notebooks may be subpoenaed in court cases that concern experimental or design questions.

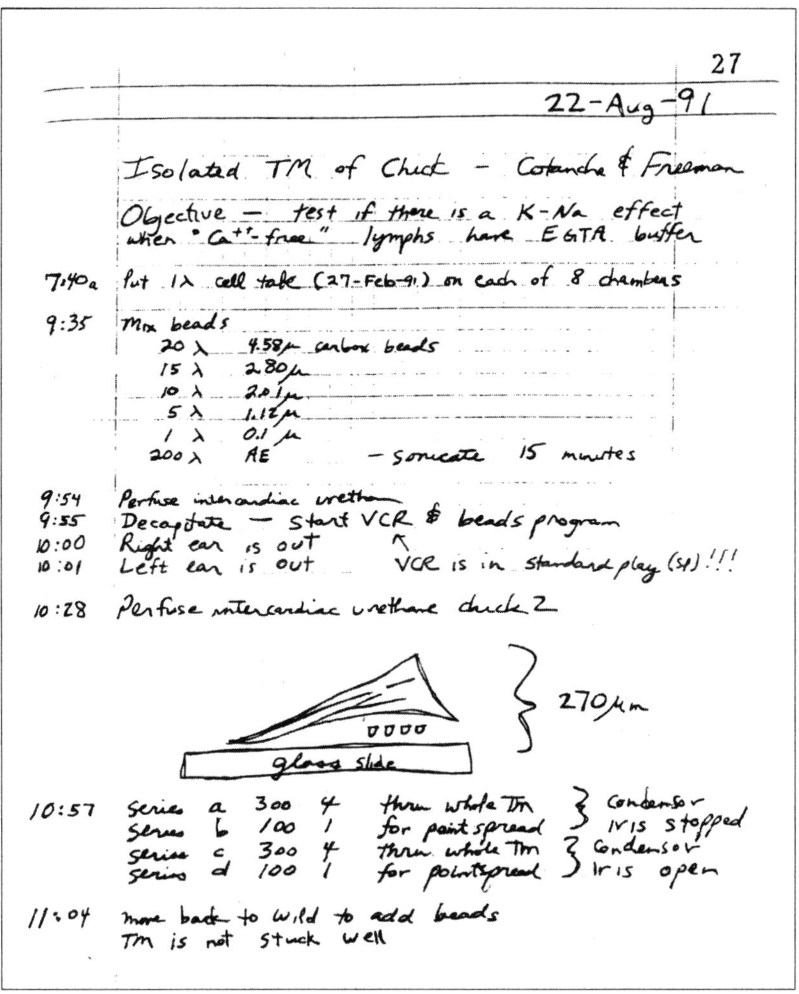
27
22-Aug-91

Isolated TM of Chick – Cotanche & Freeman

Objective – test if there is a K-Na effect when "Ca++-free" lymphs have EGTA buffer

7:40a Put 1λ cell tak (27-Feb-91) on each of 8 chambers

9:35 Mix beads
20 λ 4.58μ carbox beads
15 λ 2.80μ
10 λ 2.01μ
5 λ 1.12μ
1 λ 0.1 μ
200 λ AE – sonicate 15 minutes

9:54 Perfuse intercardiac urethan
9:55 Decapitate – Start VCR & beads program
10:00 Right ear is out
10:01 Left ear is out VCR is in standard play (SP) !!!

10:28 Perfuse intercardiac urethane chick 2

270μm
glass slide

10:57 Series a 300 4 thru whole Tm } Condensor iris stopped
Series b 100 1 for point spread
Series c 300 4 thru whole Tm } Condensor iris open
Series d 100 1 for pointspread

11:04 move back to wild to add beads
Tm is not stuck well

Figure 1.2
Notebook entries for experimental study of laboratory chick specimens. Note the statement of experimental objective and the linkage of time and action. (Courtesy of Professor Thomas F. Weiss, MIT.)

You may be liable if you fail to maintain files of calculations, sources and grades of materials, design changes, or any essential phase of your work. You may also record experimental data and notes in digital files, but, unlike laboratory notebooks, these may not be legally acceptable as original forms.

Importance of Digital Technologies

Any writing project, large or small, requires several coordinated activities that transform original data into draft documents and, finally, into finished documents. Computers help a project team tailor a body of material to fit the different aims and audiences of a proposal, procedure, memo, oral presentation, progress report, or research article.

The computer is an indispensable tool for managing the work of communication. Writers use computers for drafting and revising text, preparing graphics, searching for database and library information, communicating with coauthors, cycling documents through reviews, merging elements to produce a final document, and submitting for publication and distribution. An increasing number of documents in science and engineering are created primarily as computer files. Some funding agencies require electronic proposal submission, and many employers will accept nothing but computer-based resumes. Documents are evolving into electronic collections of knowledge from which information can be routinely assembled and reassembled in new and unpredictable ways, with new libraries of information created from selected portions of existing libraries.

A Professional Approach to Writing

The writing process is complex and abstract enough to offer many kinds of barriers. A writer who can't get started may not be able to identify the source of the problem. Inhibitors of writing are often strongly related to writing conditions such as insufficient time and constant interruptions. For many R&D writers, problems may be the result of inadequate recording and archiving strategy, confusion over the task required, or reluctance to submit draft work to supervisors and managers. You may simply not have enough information about your subject and may need to carry out more research. A writer having trouble might be stuck at any of several phases. Here are strategies to consider.

Organize Your Writing Space

Arranging your research materials and organizing your computer files can help you establish control. Anxiety over the location of materials can lead to writer's block. You're likely to need quick access to notebooks, spreadsheets, published sources, project proposals, reference works, rough drawings, note cards, and correspondence. Develop and maintain files as you work so that you can reuse information you have written.

Understand Your Task

Most communication tasks in science and engineering can be clearly defined by assessing audience, purpose, and probable formal features of the document under construction. Writers in technical fields can usually identify their initial audiences. They can shape the content of their documents to meet the information needs of a coworker, a supervisor, a reviewing agency, a journal editor, or a client. They know why they are writing, often to report or to persuade. And they are well aware of the type of communication product that is expected: perhaps a letter or a progress report or an electronic slide presentation. Writers in scientific and technical settings can usually find models for the kind of project they are working on in their own company archives.

Create a Workplan for Each Project

With an understanding of audience, purpose, and product, you can create a workplan for each project. Writing requires planning, drafting, revising, editing, and producing—activities that are usually sequential. Novice writers often equate writing with drafting and proceed without much of a plan, stringing words together into sentences and sentences into paragraphs. Beginning to write without a concrete sense of the shape of the entire document can lead to false starts, confusing introductions, inappropriate content, wordiness, and incoherent organization.

Create an outline of the proposed project and translate that outline into a table of contents that lists sections and subsections. Use what you have learned about document requirements as the basis for the sections you need to write. Think of your sections as relatively self-contained modules. With a modular plan, you do not need to write sections in order. You can begin writing a section you know well such as Methods or Results, and you can return to the Introduction or the Discussion at a later stage.

Design a Strong Visual Component

Technical documents are built of verbal and visual elements. Tables and figures are critical to exposition in all technical fields. Developing graphics early—perhaps even before you write—is an effective way to focus your work. Many writers begin organizing their work by assembling graphics and then shuffling them to work out the logical sequence of their prose. In earlier times, engineers and scientists could often expect that the tables and figures in their documents would be prepared by technical artists working from rough sketches supplied by authors. It is now almost always expected that authors will themselves prepare high-quality graphics, using electronic design and drawing tools. Plan the visual aspects of your document as carefully as you have planned for the organization and content. A powerful visual element will never emerge as an afterthought. Sensitivity to the visual aspects of technical communication even includes awareness of the final page layout and overall plan for document design.

Don't Try to Write a Perfect First Draft

A writer who expects to write a perfect first draft is likely to be the person who spends the morning putting a comma in and the afternoon taking the comma out. If you're convinced that your writing should progress routinely through a linear series of steps, you're going to hit a wall. Assume that you'll need to rewrite. Be ready to make judgments or decisions. At the writing stage, you're putting your views and findings on record. This act of formalizing can pose great difficulties when, as is often the case, your results are not all that clear. Remember that writing is itself a decision-making process. Don't put off writing until you've achieved some mythical level of certainty.

Writing and the Work of Science and Engineering

The work of science and engineering is recorded and disseminated in a variety of communication forms, both written and oral. Strong communication skills are crucial requirements for success. It is through writing that funding is secured, research processes are managed, and new knowledge is shared with others. The audiences for your writing become larger and more varied as your technical work advances from initial idea to tested final concept or product. A limited number of colleagues will have access to your memoranda, while a larger audience of peer reviewers will read your proposals. An even wider audience of sponsors will have access to your reports. When you record your findings in the form of journal publication, your contribution to knowledge will be indexed in electronic databases and available to all researchers who work on your topic.

Fortunately, working professionals in science or engineering can learn the basic principles of good technical communication as well as the special features of technical formats. In the chapters that follow, you'll find formats and strategies for a variety of writing situations. As the architect of your document, you can approach writing the way you approach other technical tasks: by understanding what information product is required; by designing a product that will meet those requirements; and by leaving time for the product testing and quality assurance that come from collaboration and review.

2

Collaborative Writing

■

As a project manager in a large R&D company, you are charged with producing a document describing a new product. The final document will span several volumes. It will include research findings, backup documentation, manufacturing plans, and quality assurance data. People from several departments, together with a few subcontractors, will need to write and review drafts. How do you begin? Chances are you'll call a meeting. With such a large team and so much to do, you'll want representatives from every department involved. Managers will need to know what data their departments are to provide. Writers will need to have their tasks defined. They'll need deadlines. The team will need to determine a review process. Knowing how to manage a complicated process may be the key to producing a complete document on time.

Writing for science and engineering is always collaborative. Whether you are a principal investigator coordinating research or a design engineer developing a new component, you need the help of others to reach your own goals. New software and hardware are constantly making group work more productive. Calendar and scheduling programs, e-mail and messaging systems, and group-authoring tools all support document distribution and revision and make collaboration easier. But managing a successful group project takes more than technology. Collaborative writing fails most often when there are misunderstandings over problem definition, research procedure, writing responsibilities, scheduling, and manuscript reviews.

Preparing Multiauthored Documents

Problems with group writing projects can be minimized by strategic planning and effective use of meetings. Authors should agree on outlines, style sheets, schedules, as well as methods for document routing, and review.

Outlines

Prepare and use outlines as control documents. An outline helps the main writer get agreement on scope and approach. Without such controls, groups are difficult to keep organized, especially when each member is producing part of a larger document. An outline facilitates the assigning of different responsibilities to different people.

Style Sheets

Establish basic formatting and documentation conventions, either by adopting a standard such as the IEEE Style Sheet or by drafting your own format for elements like subject headings, numbering systems, word usage, and figures (see sample style sheet in Figure 7.1). The simplest approach is to designate a published document or document template as a standard and use it as a guide for stylistic consistency.

Schedules

Agree on dates for milestones and set firm deadlines so that the project is not held up by straggling contributors and reviewers. Otherwise, you have no way of knowing whether your collaborators are producing their writing. The storyboarding technique described in Chapter 11 requires participants in group writing projects to pin their drafts to a wall, making adherence to schedules visible.

Document Routing

Coauthors typically collaborate by circulating documents and recording the group's comments and emendations. This review process, initiated by the main writer, may be carried out by circulating hard copy, or more likely by means of e-mail routing and annotation. Electronic routing is the handiest form of collaboration because it requires little planning and meeting time. One writer drafts the document and sends it to coworkers via e-mail. Coworkers provide critiques either by annotating the document or by directly revising it in highlighted text. Most word-processing software supports these activities.

Review and Updates

Establish a review mechanism and be sure that everyone is aware of the process. Let coauthors know what aspects of the document need to be reviewed, perhaps asking them to limit their comments to matters of technical accuracy and reserving stylistic editing decisions for a designated editor. Plan in advance for a way to ensure that updated versions of the document in progress replace earlier versions. Workgroups frequently ask the principal author or editor to be responsible for coordinating suggested changes to the document.

Holding Effective Meetings

At the start of a project, when many questions are still open, meetings are a forum for defining and reviewing problems, developing strategies, exploring methods, and critiquing results and documents. As a project progresses, meetings encourage the sharing of expertise and responsibility among colleagues, contributing to consensus building, information exchange, group decision making, and document review. But meetings can be notorious time wasters. Meetings do not easily focus attention, assume direction, or deliver concrete results. Most people see meetings as unwanted diversions. The team leader—or anyone else chairing a meeting—needs to make the effort worthwhile. To structure meetings, the chair needs to work from a clear agenda, establish effective time limits, and develop means for follow-through.

Plan the Meeting

To plan an effective meeting, you have to get everyone to agree on a meeting time and place, and you also need to inform participants about meeting length, place, and subject. If the meeting is small and informal, you can set it up over the telephone, although confirming the arrangement in e-mail is still a good idea. If your meeting involves a larger group or a formal committee, you need a written agenda announcing the place and time of the meeting, its main purpose, and the items to be discussed (see Figure 2.1). Circulate the agenda before the meeting. It is helpful if you can convince members to prepare presentations for distribution with the agenda, giving others a chance to think about complex issues in advance.

Work from a Written Agenda

An agenda should progress from (1) routine, context-setting items to (2) general information discussions that do not require decision making to (3) the main decision-making discussion to (4) recapitulation and assignments. An agenda should be of reasonable length, not so long that your meeting ends halfway through the items. The agenda, normally prepared by the person who calls the meeting, provides guidelines for conducting the meeting and keeping it on course. Generally, the chair refers to the agenda throughout the meeting, spends a given amount of time on each item, and brings the discussion to a close.

Maintain Momentum and Focus

Meetings should start and end on time, progressing so that the agenda is covered adequately. Expect to take 5 to 10 minutes to get the meeting

Memorandum

TO:
FROM:
SUBJECT:
DATE:

This meeting called by J. Aggarwal to discuss general interest in establishing a new Energy Laboratory spectroscopic facility will take place as follows:

**Thursday afternoon, December 4, from 3–4:30 pm
in Room 3-337 of the Ames Research Center.**

The main purpose of the meeting is informational. We are interested in identifying level of interest and potential sources of technical and financial support. We will determine at this meeting whether to pursue this project at the facility-wide or local levels.

Agenda:

- Introduction—J. Aggarwal
- Other national laboratories and their approaches—S. Hunt
- Some staffing and computing requirements—L. Dickson
- Possible local sources of technical and financial support—J. Aggarwal

Discussion:

- How to proceed in the next phase
  1. Should we handle this as a facility-wide or as a local laboratory need?
  2. Should this need be addressed next year, or should we wait to see what happens in our sponsored funding patterns?
- Other items for future agenda

The discussions will be preliminary and the only decisions we will make at this meeting will be to answer the two questions concerning proceeding in the next phase.

Figure 2.1
Sample meeting agenda. Note the combination of assigned presentations and decision-making discussions. Detailed agendas improve participation by providing participants with a chance to think about issues in advance.

under way and to frame the discussion. Time the meeting, allotting each item so many minutes, and then move on. Digressions quickly wreck meetings. Effective meeting dynamics require that the chair view each discussion as part of a whole and move the group on at a steady pace.

Routinely restate the topic and remind participants of the issue under discussion. People will readily stray into other topics, some important and some irrelevant. The chair—or even interested participants—should bring a straying discussion back to the agenda. If participants want to move into new productive territory, reserve time for the topic at the next meeting or allow discussion at the end of the meeting. You can do this diplomatically by appealing to time and agenda constraints.

Keep Participation High

Encourage everyone to engage in the meeting. Don't allow ten people to attend a meeting where three participate while seven others sit and say little or nothing. The result may be narrow use of available expertise and loss of consensus.

Promote participation by studying and assessing personalities and work styles. Call on silent participants and neutralize excessive talkers diplomatically. One effective technique is the roundtable query, in which every member of the group is asked to respond to a question. Also effective is calling on those with specific expertise to prepare brief presentations.

Establish a Record

Memories of discussions soon fade, and entire meetings can be consigned to oblivion because no one has jotted down a record of the main discussion points and the decisions reached. Taking notes during or immediately after the meeting establishes a record of ideas, names, and agreements. If the meeting is informal, each member might record notes in a personal notebook. If the meeting is formal, the chair needs to designate an administrator who will take notes and prepare minutes for later reference.

Minutes should summarize the main points discussed and the decisions made. Only occasionally is precise wording necessary, and then only if the wording is important. Minutes are not normally verbatim records; tape recorders do that better. The minutes for an hour-long meeting, for example, will not normally run more than one to one-and-a-half pages. Minutes need to be submitted for the approval of the participants.

Monitor and Promote Follow-Through

Perhaps the most difficult meeting task is to translate decisions and commitments into concrete actions. Every meeting decision should be written down and also followed by a discussion of the means and schedule for the project. Many organizations circulate a list of Action Items immediately after meetings, reminding participants of agreements and deadlines (see Figure 2.2).

Supervisory Collaboration

Collaboration can also take a hierarchical form. In some organizational settings, supervisors review the writer's work both for its technical accuracy and for its institutional implications. Supervisors typically review assertions and recommendations for the way they reflect the policies of the workgroup and the larger organization.

Seniority or authority characterizes supervisory collaboration, as the supervisor can *require* the writer to make certain changes. For example, a report assessing how effectively a contractor is meeting the terms of an agreement might contain much criticism. The writer may feel that the critique is justified, whereas the supervisor may feel that the criticism is harsh and antagonistic. Differences in perception are common to all collaborative writing, but a hierarchical relationship can make collaboration potentially abrasive.

Once again, meetings are an effective means of discussing differences of opinion and reaching a preliminary understanding. In supervisory reviews, all parties benefit from early agreement, before the writing has proceeded so far that the writer has trouble carrying out the revisions. Collaboration is much more effective when the parties achieve early agreement because, as writing progresses, the writer invests more time and identifies more intensely with the work. Personal ego is increasingly

IEEE Professional Communication Society
Administrative Committee Meeting, New Orleans, LA
7 September 2001

ACTION ITEMS

Item 1.

Description:	*Prepare ad for Computer Society publications.*
By Whom:	*Beth Moeller and Leann Kostek.*
When:	*1 October 2001.*

Item 2.

Description:	*Locate funding sources within IEEE for video project.*
By Whom:	*Roger Grice.*
When:	*1 January 2002.*

Figure 2.2
Action items distributed to meeting participants, reminding them of what they have agreed to do. Many workgroups routinely ask one member to prepare a list of action items and to distribute it through e-mail.

at stake, and required revisions become harder to swallow. The object is to avoid situations in which a writer thinks a document is finished but must then extensively revise it.

The New Technologies of Collaboration

Electronic collaboration technologies now enable group members to schedule meetings, share information, monitor progress, and review documents. Many teams have adopted e-mail and electronic discussion forums as their primary forms of communication. Distance and time are increasingly irrelevant. Team members in the same or distant locations can read documents in progress and contribute their comments off-line. Or they can collaborate in real time through scheduled network meetings, actively sharing electronic whiteboards, discussing and editing displayed documents.

Group members assigned the task of taking minutes can enter their records on laptop computers. They save time by creating in advance a customized template for the meeting records, based on the agenda. Newer note-taking applications allow for the creation of multimedia meeting records that become part of the workgroup's shared resources. Recorders combine personal notes, presentation slides, or other material in a single, unifying electronic document, and they share that document with an entire work group via the Web.

Michael Dertouzos, the late Director of the MIT Laboratory for Computer Science, was confident that we will soon have an intelligent advanced authoring tool for meetings (ATM). Using the ATM, the person assigned the task of note taking builds a structured hyperoutline in advance. As participants speak, the note taker hits different keys on a computer keyboard to record pivotal statements under one of several categories of discussion already set up. Speaker identification is electronically accomplished through computer analysis of voice samples. The spoken fragments are also directed to a speech-understanding program, where they are transcribed and indexed, as well as summarized. The ATM also records material displayed by meeting participants. Any member with access to the group's deliberations can dial into the hypersummary and receive, in answer to a query, an audio version of key statements on the topic, an on-line text version, as well as associated slides and supporting material.

Guidelines for Virtual Meetings

Network-meeting software enables virtual interactions in real time. Parties to the meeting participate in chat sessions and also share and collaborate on documents. Effective on-line meetings require adherence to a set of practices so that the meeting can be productive and fair to all participants. Many of the guidelines for effective virtual meetings are identical to those for face-to-face sessions: The agenda and all relevant documents should be distributed in advance, and the moderator of the online meeting needs to keep the meeting on track and ensure that all participants are contributing.

Other guidelines are responses to the unique on-line environment, where slow typists are at a disadvantage and fatigue is a crucial factor for all participants who are watching responses on the computer monitor. Discussions in an on-line meeting are much slower than in a face-to-face meeting, and participants can be confused about whether others are still present and attentive.

On-line meetings are more effective when participants have received guidelines in advance of the meeting and have agreed to adhere to them. Ask participants to set up a split screen so that the agenda for the meeting stays visible in the left screen and other information is on the right. Tell participants to "speak" to others when they first enter the virtual meeting so that everyone else knows who is present. Require those who need to leave an on-line meeting for a brief interval to note when they are leaving and also when they return. Agree on a set of on-line meeting typographical conventions such as HU for "hand up," indicating that you want to ask a question or an ellipsis (3 periods) indicating that you have more to say on a topic.

Collaboration in Context

Collaboration is essential to research and writing in technical fields. Knowledge is advanced through teamwork, and researchers contribute their expertise to group-authored documents. Advances in computer technology have improved support for collaborative work—beyond the constraints of face-to-face meetings and inefficient note taking. But effective collaboration requires more than hardware and software. It requires a willingness to negotiate the difficulties of working with others, remembering that successful teamwork so often yields richer interpretations and stronger arguments. In the following chapters on proposals, reports, and journal articles, you'll find strategies and options for managing time, tasks, and people as you produce complex documents.

3

Your Audience and Aims

■

Research scientists studying how brain neurons fire have to write grant proposals for project support and eventually publish papers for colleagues. But neuronal firing means different things to potential funders and colleagues. The consulting civil engineer preparing a report on soil samples at a bridge site is writing for architects, building contractors, town managers, and Environmental Protection Agency agents. To be effective, both the researchers and the engineer must consider the audience, the people who will read their writing.

Identifying the readers' needs and interests turns out to be one of the most important parts of writing. Science and engineering are problem oriented, and stating problems clearly helps focus resources on answerable questions. To keep problems from existing in purely abstract terms, a writer needs to identify the audience interested in the problem.

The question then becomes this: What is the best strategy for meeting those readers' needs? For example, an industrial engineer might see automating a manufacturing operation as a technical problem. But it is also a financial problem that needs to be justified administratively by managerial decision makers. The writer whose proposal simply concentrates on a technical explanation fails to shape the arguments for the readers who will ultimately make the decision. This misunderstanding can defeat a writer's aim.

Your Readers' Interests

Readers are usually motivated by their job responsibilities as decision makers (managers), knowledge producers (experts), operators and maintainers (technicians), and generalists (laypeople). But these different audiences are abstractions or, at best, averages. Not every expert in particle physics is going to think the same way, use the same methods, or have the same problems. The veteran technician knows more about many technical subjects than the university-trained colleague.

Addressing your audience is even more complicated when the audience includes managers, specialists, technicians, and laypeople. Each part of your audience will need to find the information it needs. An audience of managerial readers, for example, will evaluate what you have to say in the terms of their decision making: costs, benefits, alternatives. The expert, technician, and lay reader will also analyze your message according to his or her interests and responsibilities.

Some documents have a primary audience, which you can often identify by clearly defining the purpose of the document. For example, if you aim to establish a new procedure for preparing titanium dioxide, a commercial white pigment, by method X, then you are addressing individuals with technical concerns. If, on the other hand, you aim primarily to show that titanium dioxide precipitates prepared by method X are 30 percent more durable than those prepared by method Y, then you are speaking to experts interested in innovations. If instead you set out to argue the feasibility and economy of a three-year $800K program to develop an industrial process for synthesizing titanium dioxide by method X, you are writing for managers concerned with planning and resource allocation. Your aim should identify your audience.

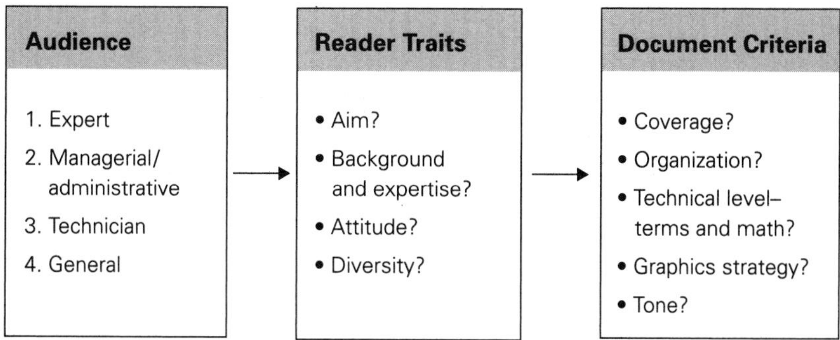

Figure 3.1
The interplay of audience, reader traits, and document criteria. Reader needs determine document criteria.

Coverage, Organization, and Technical Level

Producing writing appropriate for a given audience cannot be done by formula. Instead, you need to work with several variables as you shape a document for a projected readership. The expectations of your audience should determine the coverage you give your subject, the organization you give your material, its technical level (including graphics), and, finally, your tone (Figure 3.1).

Choice of material is your first decision. Coverage refers to the scope of the subject, and it can vary greatly. Let's say that you are planning a major three-year program to develop a new speech recognition process for improving on the human-computer interface. You can cover your subject in many ways, depending on your readers' main concerns. Someone needs to finance and administratively support the work; others have to be convinced of the feasibility of the process and agree with your breakdown of the technical elements of the project. For an audience of managers or entrepreneurs—those who must decide whether your project merits funding—you might focus on the technical feasibility and commercial advantages of a new process for speech recognition. For technical experts, you might more fully describe the technical aspects of the new speech recognition process. Even the lay reader gets into the picture sometimes as a general client, entrepreneur, or local citizen. Your focus may accordingly shift toward speech recognition applications in business, medicine, and so on. It's up to the writer to work out who the audience is and what they need to know.

Organize your material around your readers' priorities. For example, assume that you are a civil engineer writing to analyze the seismic site conditions of a proposed San Francisco harbor facility and to make appropriate design recommendations. You can organize the same information in two distinctly different ways, each anticipating the priorities of the specialist or the managerial reader, as shown in Figure 3.2. The first column develops materials so that an expert can analyze the process leading up to the design recommendations. This kind of organization

Topic: Seismic Design of San Francisco Waterfront Facilities

Expert audience	**Managerial audience**
1 Introduction (problem and background)	1 Summary of problem and findings
2 Analysis	2 Design recommendations
2.1 Local geology and seismicity	2.1 Background
2.2 Site subsurface conditions	2.2 Recommendations
2.3 Potential seismic hazards	2.3 Discussion of design recommendations
3 Experimental methods	3 Analysis
3.1 Slope stability tests	3.1 Local geology and seismicity
3.1.1 Methodology	3.2 Site subsurface conditions
3.1.2 Results	3.3 Potential seismic hazards
3.2 Earthquake-induced yard settlement tests	3.4 Discussion of experimental results
3.2.1 Methodology	4 Conclusions
3.2.2 Results	**Appendix A** Data on slope stability
4 Discussion	**Appendix B** Data on earthquake-induced yard settlement
5 Design recommendations	

Figure 3.2
How the same material (on seismic conditions in San Francisco) could be organized for two different audiences.

favors critical evaluation by a knowledgeable specialist. The second column emphasizes the major findings and recommendations that would be most important for a manager-planner. Conclusions, therefore, come first, with details given in order of likely interest. The detailed experimental information is included at the end, to support the recommendations.

A third way to tailor documents to audiences is to adjust the language, especially the special terminology, mathematics, symbols, and graphics. Assume, for example, that you want to explain the operation and use of the carbon dioxide laser scalpel. The subject deals with laser technology applied to surgery. The topic is specialized in both engineering and medicine. The audience might include hospital administrators, surgeons, patients, nurses, medical technologists, research engineers and marketing personnel. Considering their needs and educational backgrounds should help you determine what language is appropriate.

The following passage might be addressed to research engineers. Its fairly advanced technical level is reflected in the terminology and use of symbols, references, and units:

> The CO_2 laser scalpel connects to a monotoxic optical fiber made of silver halide, originally constructed and tested at the University of Tokyo (Atsumi et al. 1983). The core of light fiber that transmits the infrared beam of the CO_2 laser is made of $AgBr_2$ and is clad with a layer of $AgCl_2$. Transmission loss in the fiber is 0.22 dB/m at 10.6 mm.

Specifying output and other operating parameters gives the reader technical information about the instrument's capabilities. References to the literature enable the user to obtain other important data. This level of technical discourse would be inappropriate for nonexpert audiences, who could not be expected to understand the language.

The same topic directed toward the surgical nurse would incorporate less technical language but would cover the operating procedures. The emphasis here is on safety and standard practice, with directly worded prose.

> Although the type of laser used and the surgical applications are determined by the surgeon, the nursing staff must ensure that the equipment is regularly inspected and maintained and that potential fire hazards are avoided during operation. The assistant should drape the area adjacent to the operative field with wet towels, which should be remoistened frequently during the surgical procedure (see Figure 6). A large container of saline solution must be kept available, both to moisten the operative area and to douse flames if material ignites.

The technical level of the language is also influenced by the accompanying graphics, which can vary greatly for any given subject. The graphical presentation of information can range from highly specialized line graphs to general pictures.

Document Pathways

Another way of thinking about audience is the document pathway. You know that your report or memorandum travels, accumulating readers along the way. Most organizations have a hierarchy through which written communication passes. The organizational chart often shows some of this path. The document may well move up the hierarchy through supervisor 1, the group leader, to the division manager, and finally to the research director. Then it might travel back down to the staff level of some related group or over to marketing. Figuring out where the document will go is in a writer's interest. The document pathway will tell a lot about your audience and therefore about the general coverage and organization needed in your document.

Your understanding of this audience comes from knowing what the people reading your document will want to do. Some will work with you on the draft to improve the content and presentation. Others will read the finished version and take action. Still others may skim and file it, and many will just dump it in the wastebasket. Some may even use it to assess the quality of your work. The best preparation you can make for understanding your readers is to study how your company or institution is organized and familiarize yourself with its staff and their responsibilities.

The Peer Specialist. Assume, for example, that you are part of a design group reporting on the structural and commercial advantages of a new lightweight composite for building installation frames for solar panels. Several structural engineers and materials specialists might comment on your first draft. These group members share an interest in the success of the report. They read the document, add their comments in the margins, and contribute ideas. One or more of them might be listed as coauthors. As an audience, then, this group is closest to the subject matter and may be the most technically informed.

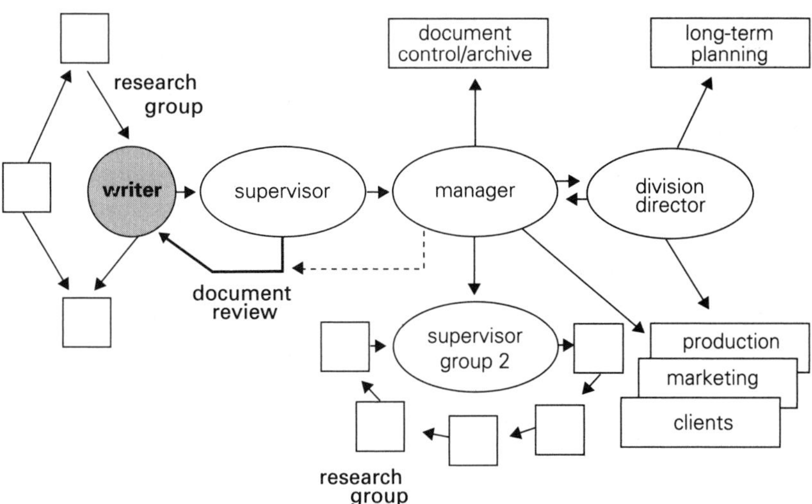

Figure 3.3
The R&D writer's rhetorical world. A document typically finds its way into several different situations in which individuals of the local organization use the information in different ways.

The Supervisor. The group supervisor, who typically is held accountable for meeting larger corporate research objectives, has a major stake in the documents the group produces. The supervisor wants to make the document focus on established objectives in an effective way. Draft documents often fail to make a clear statement. They also fail, sometimes, to address the established aim of the group enterprise, or they don't communicate in effective and organized prose. This person may comment on the document at its first draft stage and cycle the document back for revisions (Figure 3.3). Reading and revising can be very demanding, especially for novice writers. As documents return for second or third revision, the tension can build.

The Manager. R&D groups often report to a manager, who coordinates and directs the broad effort. This higher-level administrator does not participate, normally, in preparing individual documents but is undoubtedly one of the document's most important readers. Like group supervisors, managers release documents to meet their responsibilities for knowledge production, but they also monitor research goals and progress, a process in which the written report plays an important part.

Writing for Publication

Specialist, manager, and lay audience take on new meanings when you write for the published record. Away from the in-house R&D environment, the audience often becomes easier to identify. The large, specialized readership of a journal, for example, can be assumed to have similar education, professional interests, and technical expertise. Most articles are written to expand the knowledge base in a given field, and an expert audience can therefore follow terminology, mathematics, and methods. You can develop communication strategies partly by inspecting published documents (e.g., proposals, refereed articles, formal research reports). The coverage, language, and graphics follow certain conventions throughout a journal publication.

The document pathway typically moves through an editor to several referees (Figure 3.4). The editor is not your supervisor but the employee of another organization. Note that Figure 3.4 shows two cycles. The editor acts as coordinator and stylistic overseer for one cycle, and referees judge the document's technical merits in the other cycle. If the document in question is a proposal, this outside referee audience will file

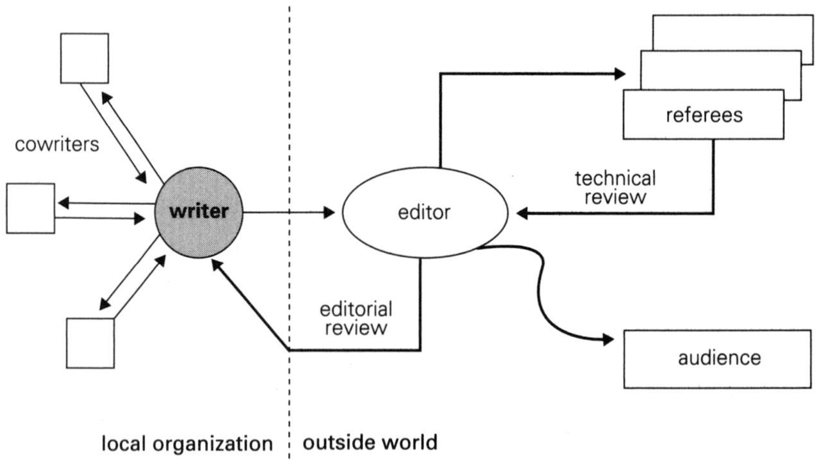

Figure 3.4
The rhetorical world of a writer contributing to a refereed journal. Although the review process is similar to that followed by the R&D writer, the journal writer's review is more formal and moves outside the local environment.

reports supporting or rejecting the funding. If the document is a refereed article, the referees file reports supporting or opposing publication. Often, the referees recommend revision followed by resubmission.

Framing Writing Projects

Science and engineering are part problem solving. Defining your problem will improve your writing. The more specific your problem, the better you can design your research and determine your goals.

But problems can be approached in many different ways, depending on your audience. Developing smart air bags for the automobile industry poses different kinds of problems for the CEO of General Motors, the head of its marketing division, designers in its air-bag development and testing teams, federal regulators, and the car owner. To frame a problem you have to identify your audience and refine your topic.

From Technical Problem to Writing Topic

A problem is a conflict that someone wants to remedy. For example, "the scanner on board the *Landsat* 4 satellite is malfunctioning" is a problem. A topic renders the problem and its solution into a focused research and writing objective. Some possible topics related to the satellite scanner malfunction are

- Causes of scanner malfunctioning on *Landsat* 4
- A protocol for correcting *Landsat* 4 scanner malfunctioning
- An improved design for *Landsat* electromechanical scanners

Each topic reflects the original problem but refines and limits it for an audience. Narrowing your topic in this way defines your research and writing goals.

Your topic reflects your audience and its interests. In research environments, problems typically reflect state-of-the-art theory and practice and hence are more abstract and specialized. For example:

- What is the effect of single amino acid replacement on the thermal stability of the N-terminal domain of a k-repressor?
- How can we characterize nematic ordering in lyotropic liquid crystals?

• What are the effects of iron additives on soot particle formation and growth?

Here problem definition is demanding. It takes a seasoned research engineer or scientist to identify critical questions that lie within the domain of experimental or theoretical possibility.

Engineers and scientists working in industrial R&D generally don't generate the topics they write about. They work on problems identified by clients, colleagues, research directors, or supervisors. For example:

• What is causing the recent chemical degradation of our O-ring seals?
• What blade design should we use in our C-53x windmill?
• Should we develop a mechanical suspension system or air cushions to improve the ride quality of our XX-100 vehicle?
• What fire-alarm system will best protect our client's warehouse installation?

Yet even these concrete problems require further refinement before you have a topic you can confidently start to work on.

Limiting Your Topic

Limiting your topic increases your control over the subject matter. By narrowing your goals to what you can do within your resources, you improve your chances of success. For example, you would not want to write about an improved design for *Landsat* scanners if you had only studied the causes of their malfunction. Narrowing the topic helps you avoid committing yourself to a project with no boundaries.

To limit your topic, make it more specific. Read background material and related project work and discuss these with colleagues, experts, clients, or supervisors. Write down what you know about the problem you're addressing. A clear problem statement, such as "How can we reduce the spread of Lyme disease in public parks?" helps focus your investigation. Then limit your main topic. A topic like "Lyme disease in public parks" is much less manageable than one like "Recommended measures for reducing Lyme disease at Crane's Beach." The former topic is broad and requires a wealth of information to cover; the latter is specific and restricts the scope of your research to a local effort. Refining your topic saves you time at the writing stage.

General	*Specific*
Lyme disease in public parks	Recommended measures for reducing Lyme disease at Crane's Beach
Improving the uniformity of titanium oxide ceramics	Synthesizing uniform titanium oxide precipitates by the ethanol-water method
Proposal to study new methods for treating alcoholism	One- and two-year effects on alcoholics treated by individualized behavior therapy

Avoid becoming rigidly attached to your first problem definition. Initial assumptions are often quite wrong. Problems are likely to evolve with information gathering. Experiments, reflection, and discussion with colleagues will help you consider alternatives. Adams notes in his study of problem solving (1974):

> Much thinking went into the mechanical design of various types of prototype tomato pickers before someone decided that the real problem was not in optimizing these designs but rather in the susceptibility of the tomatoes to damage during picking. Part of the answer was a new strain of tomatoes with tougher skins.

From Topic to Aim: The Goal of Your Document

A document is a slice of your work, one that needs its own structure. Before you begin drafting, therefore, you need to develop an aim for your document. An aim is the reason for writing the document, a specific goal. When you develop a research topic into a document aim, you convert a category to an operation. You propose to do something for someone, namely your audience. Faced with a scanner malfunction aboard *Landsat* 4, for example, you might aim to demonstrate that *Landsat* 4 malfunctioned because a specific part in the scanner assembly jammed against the assembly mount. Or your aim might be to describe a detailed protocol for remotely manipulating the motor of the scanner assembly.

Defining your aim means asking yourself, Why am I writing this document? Usually, your answer to this question tells you that you want to describe physical objects and processes, narrate developments or analyze your topic, in order to get your readers to do something or persuade them of something. Even specialized topics can be approached with different aims, as the following example shows:

Topic: Effects on the U.S. biotechnology industry of federal guidelines for genetic research.

• *Description*: To describe federal guidelines for classifying laboratories for genetic research.

• *Narration*: To provide a case history of the way the LMN, Inc., cloning effort was influenced by federal guidelines for genetic research.

• *Analysis*: To compare the effects of centralized regulation on industrial research costs in insulin synthesis programs in the United States and Great Britain.

• *Persuasion*: To show that federal guidelines have not placed the U.S. biotechnology industry at a disadvantage with those in Europe and Japan.

As you define your problem, identify your aim, and home in on your audience, you create the framework for your writing. One way of pulling all these considerations together is to write a paragraph or two that reflect your thinking, a statement of aim. This statement defines your audience, refines your topic, and focuses on a specific argument. Consider the following project:

In three months of work on a control algorithm for remotely adjusting space vehicle direction and speed, two design engineers gather a dozen memoranda and design documents on guidance control strategies. They maintain two laptop files with observations and calculations treating space vehicle design and a new selection algorithm for firing space vehicle jets. They fill another file with the minutes of meetings at their home research organization and the contracting government organization. They also develop several large digital files of computer simulations, engineering drawings, rough schematics, tables, and graphs. The research, carried out at a national aerospace laboratory, must be condensed into a report 30 to 40 pages long, including flow charts and the recommended selection algorithm.

For these design engineers, a statement of aim expresses their intentions. Like any statement of aim, theirs should answer three questions:

1. *What is the primary aim of your document?* In the first few sentences, try to state your goal in simple operational language that implies action. If the problem is controlling the translational velocity and orientation of the vehicle, you might start out by saying that you will outline a method for controlling the translational velocity and orientation of the space vehicle by establishing a protocol for initiating firing times for a group of jet thrusters.

2. *What problem is being addressed?* After you have established your aim, examine the problem in more detail by establishing the situation

and conflict you are addressing. You might cite previous work by others on the problem. A problem usually consists of several parts or key variables. By identifying what these variables are, you establish your perspective on the problem. You also tell your readers just how you are going to treat your material.

3. *What is the scope of your document?* By stating your objectives in the document, you provide the kernel of your argument. Keep the objectives simple but specific enough for your reader to grasp your method of solving the problem. The objectives determined by our design engineers, for example, could argue that the solution is to adjust jet-firing times by adopting a selection algorithm that minimizes the errors in the jet-firing times.

Aims Imply Audience

The key to drafting a statement of aim is to keep it simple and operational. You strip away most of the qualifying detail to arrive at your central goal, the problem addressed, and your specific objectives. The process, although always difficult, forces you to come to terms with the priorities for your work. It is a process of clarification.

Readers read to solve their own problems. If you state your aim clearly, your potential readers may make informed choices about whether or not to read your work. If you stick to your aim throughout your document, a reader who shares your aim will keep reading. Always ask yourself, What are my intended readers going to do or know after reading my work? Keep this question in front of you as you draft your document.

4

Organizing and Drafting Documents

■

You have a research problem to write up. You've spent weeks on the solution. Your colleagues agree that you've found the right approach. You've thought about your audience. You begin to see on paper what exactly you can—and cannot—claim. Your discoveries are about to assume their soon-to-be-transmitted shapes.

Your impulse may now be to sit at your word processor and write your document, from the first page to the last. Chances are that won't work. You still need to think about organization. You will need an outline. Outlining is a powerful means of analysis and synthesis, a tool that helps you develop your prose strategy.

There is no standard way to outline. All outlining is a process of trial and error. Some people work with crude scratch outlines. Others use formal patterns. Still others use templates from word-processing packages, which can help organize material. No program, however, is a substitute for logical thinking. You can't outline merely by following a formula.

Outlining as Organization

The process of outlining partitions your document. You divide your materials under topics, sequence the topics, and then further subdivide them into subordinate ideas. As you arrange topics, you also mold a structure of key points that shapes your work. Outlining effectively isolates and sequences the categories of interest. As you explore the relationships among these categories, you define the limits and emphasis of your document.

Think of outlining as a stage of the writing process. It can help you in the following ways:

• *Isolating and arranging topics.* As you name some topics and dispense with others, you give focus to your document. To develop the parts of an outline, identify keywords that name your categories and reflect your aims. As you list these words and phrases, you can also consider subcategories and explore the logic of your source material. For example, an investigator studying how fires damage concrete walls that have been repaired with epoxy might start with a loose listing of topics, as seen in Figure 4.1. This initial topic list or scratch outline partitions the subject. It also helps the writer set priorities and think about connections among topics. After more thought, the writer might partition the topics further, as seen in Figure 4.1.

This exploratory process eventually leads you to connect the parts of your document logically and to fill in the gaps. Often, however, you have to reconcile this conceptual logic with the standard formats required by journals and funding agencies. These formats, some of which are shown in Figure 4.2, are conventions that assist readers by organizing material in predictable ways.

• *Integrating a general document design with your specific material.* In designing a document, try to integrate a general format with the specifics of your project topics. For example, the case study format seen in Figure 4.2 provides a general conceptual design for the specific topics of a failure mode analysis report, as seen in Figure 4.3. Your general format may be the standard requirement of a funding agency or government specification. It may also be your own top-down design of a series of categories. As you structure the document, keep integrating the general format with the more detailed topic outline.

• *Adjusting the scope and sequence of your source material to reflect the needs of your audience.* Outlining is a means of data reduction. As you arrange information, eliminate what you do not need. Two common

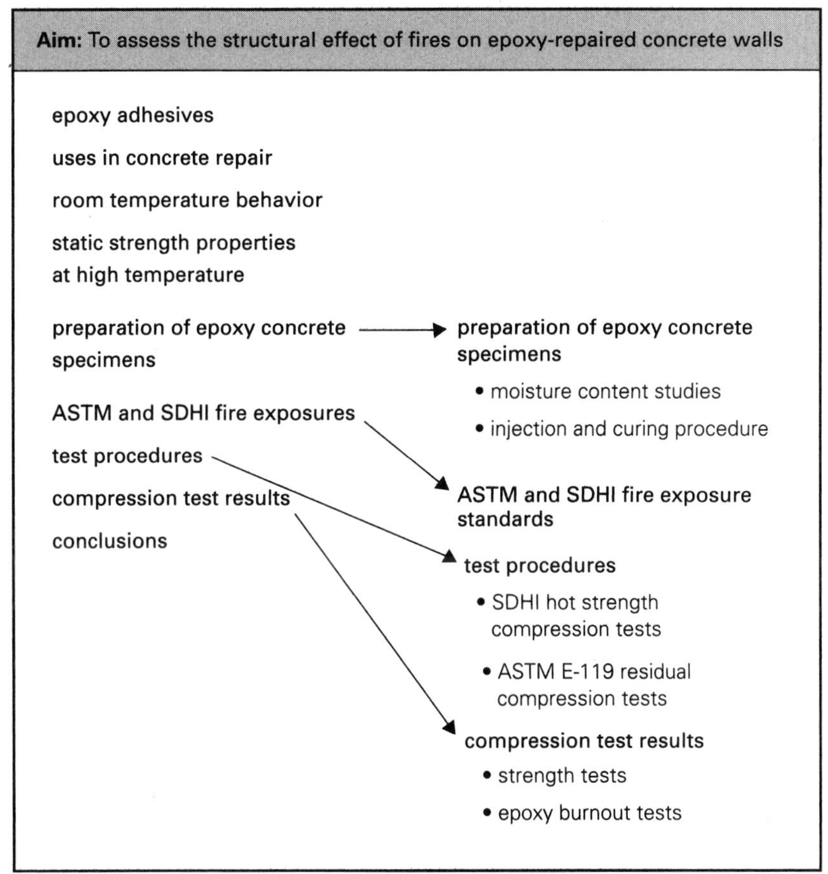

Figure 4.1
Outlining as a process of partitioning your subject by listing topics. Such topic development helps you set priorities and think about connections among topics.

approaches to sequencing information, as seen in Figure 3.2, are logical order and order of importance. Arrangement in logical order is more likely to appeal to experts, whose interests are conceptual; following the order of importance is more likely to appeal to managerial and administrative audiences concerned with costs, personnel, and schedules.

• *Getting feedback from colleagues and supervisors.* One important use of outlining is to promote consensus on goals, coverage, and strategy for documents, whether proposals, articles, or theses. Circulate your working outline for discussion. At the outlining stage, the suggestions of a collaborator or supervisor may save you a lot of time later in the writing process. Changing outlines is much easier than changing drafts.

Some typical document formats **Journal article**	**Formal Report**
[Title, abstract]	[Title page, executive summary, table of contents, list of figures, list of tables, nomenclature]
Introduction • Aim, problem, scope Theoretical section Experimental section Results Discussion Conclusion	Introduction • Aim, problem, scope Theory Experimental section Results Discussion Conclusion Recommendations
Acknowledgements References	References Appendixes
Case Study	**Memorandum, Lab Report**
[Title, abstract,table of contents, list of figures, list of tables, nomenclature]	Salutation [To:, From:, Subject:, Date:]
Introduction Aim, problem, scope Procedure Case 1 Description Analysis Results Case 2 Description Analysis Results Case N. Description Analysis Results Compararive analysis Conclusions	Problem and background Discussion and conclusions Experimental procedure

Figure 4.2
Some typical document formats. Such structures are top-down design tools that serve to reduce materials into predictable patterns for readers. These formats also represent logical structures appropriate to specific kinds of subjects, methods, and data.

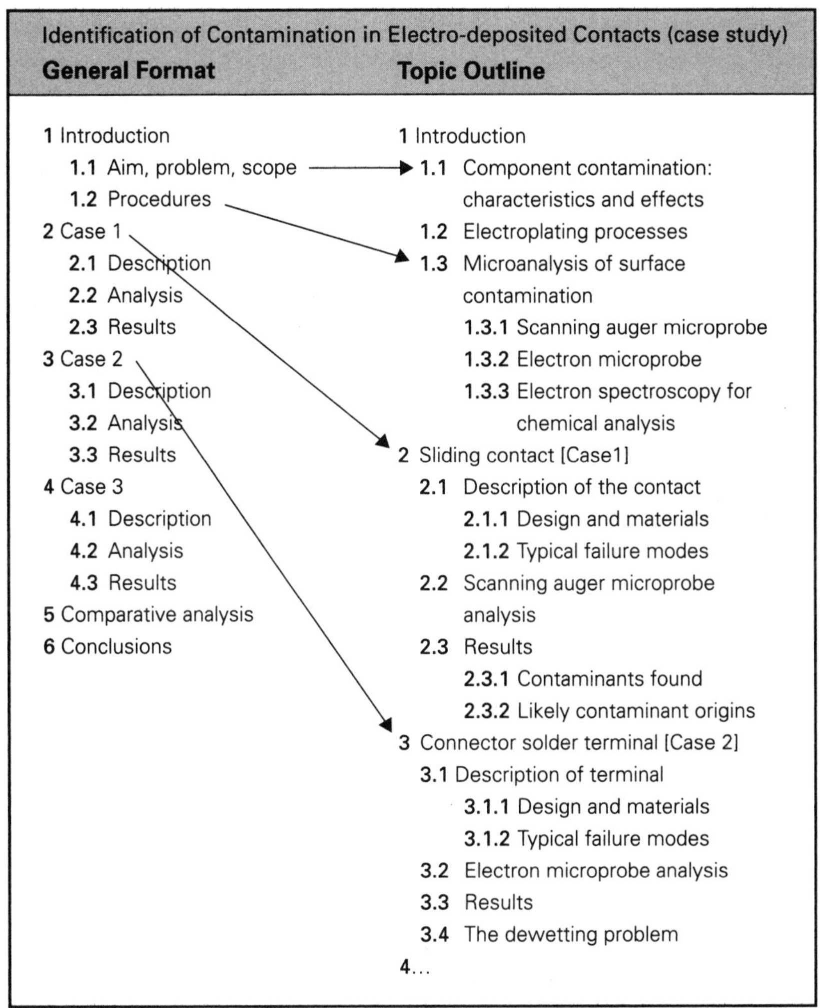

Identification of Contamination in Electro-deposited Contacts (case study)
General Format
Topic Outline
1 Introduction
1.1 Aim, problem, scope
1.2 Procedures
2 Case 1
2.1 Description
2.2 Analysis
2.3 Results
3 Case 2
3.1 Description
3.2 Analysis
3.3 Results
4 Case 3
4.1 Description
4.2 Analysis
4.3 Results
5 Comparative analysis
6 Conclusions
1 Introduction
1.1 Component contamination: characteristics and effects
1.2 Electroplating processes
1.3 Microanalysis of surface contamination
1.3.1 Scanning auger microprobe
1.3.2 Electron microprobe
1.3.3 Electron spectroscopy for chemical analysis
2 Sliding contact [Case1]
2.1 Description of the contact
2.1.1 Design and materials
2.1.2 Typical failure modes
2.2 Scanning auger microprobe analysis
2.3 Results
2.3.1 Contaminants found
2.3.2 Likely contaminant origins
3 Connector solder terminal [Case 2]
3.1 Description of terminal
3.1.1 Design and materials
3.1.2 Typical failure modes
3.2 Electron microprobe analysis
3.3 Results
3.4 The dewetting problem
4...

Figure 4.3
Converting a general case study format to a specific topic outline. Outlines effectively reduce data by partitioning material and creating subject focus and scope. They also show topic sequence and logical organization. In addition, they serve as writing and revising aids, as well as furnish section headings for document design.

Drafting the Document

Drafting is never the same for any two writers. Methodical writers work from an outline, point by point. Intuitive writers may write entire sections at a time, barely glancing over their outlines. Outline are maps that can keep you on course and remind you of your audience and aim. You may want to add or delete topics as you write. Expanding on an outline in a word processor is one way to create a rough draft. For example, an outline entry such as "moisture in concrete samples" can become an assertion:

Because moisture is never entirely absent from concrete, simulated fire exposure studies should factor in a heat absorption capacity for an amount of water 2.5 percent of the total sample weight.

With such claim statements, you expand your main arguments outward from the outlining stage.

Tools and Tips for a First Draft

Word-processing programs are effective tools for merging the outlining and drafting processes. You can develop an initial outline, start writing individual sections of it, and then rework material that needs further development. Your outline can be fleshed out and your draft revised in whatever order you feel comfortable with. Some writers type the draft into a word processor; others write it in longhand, have it word-processed, revise on a printout, and have it word-processed again.

Here are some tips for drafting:

• *Review your aim.* Try to keep your main writing goals in view and avoid digressing.

• *Set writing goals.* Writers write drafts most successfully in stages. Set an objective for a four- or five-page section and write the section at one sitting, if possible.

• *Maintain momentum.* Keep writing. Don't insist on achieving finished copy. Don't worry about where you start. You can begin with the concluding section and end with the introduction; that way, your conclusions will be focused on your introductory claims. By writing out of document sequence, you can build up writing momentum for more difficult sections.

- *Revise rigorously in hard copy.* Computing encourages wordiness. If you print out the text, you can get a sense of the whole, and you can jump around quickly as you edit for organization and consistency.
- *Expect to go through several draft and printout cycles.* Use the revision capabilities of computing to review and revise your draft. Don't put all your time into your first draft.

The very first draft, usually called the rough draft, is something you generally don't show to anyone. Few people other than you will be able to read it intelligently. A rough draft is useful, if messy, because it allows you to organize your document and work out your main arguments. Your next draft may still be crude, but it may be ready for the attention of others.

Cycling a draft can help you immensely if it brings constructive criticism. Cycling is a requirement in many organizations, but you should follow one firm rule: Don't submit crude drafts to colleagues or supervisors. Crude drafts often get treated as final drafts. A crude draft may be nearly complete, but you don't really have a finished draft until you have the following:

- An introduction, middle development, and conclusion, so that coverage and analysis can be examined
- Coherent, grammatical language
- Accurate spelling
- A series of clear, if still rough, graphics
- Clean copy, with subject headings, subheadings, and standard margins

Be sure you've reached this stage before you begin to circulate your copy to those who will judge your work. Readers soon forget the distinction between rough and final copy, and they may associate your performance with what they see first.

Developing Arguments

An argument supports a claim with a convincing set of reasons. You can expect to make arguments in nearly all documents, whether technical reports, refereed journal articles, or memoranda. A writer's aim is rarely just to inform. You need to identify problems, make claims, and defend them convincingly. Generally, the facts, no matter how effectively you analyze them, will not unequivocally support your claim. Writers need to interpret the facts to show that they mean what the writer thinks they mean.

Facts rarely speak for themselves. Readers need to understand the context of the facts. They must be convinced that the facts are accurate and effectively used. Hence, readers need to know exactly what your claim is, what problem you think you are addressing, what you think are the issues of the argument, how you got your evidence, and what you think your evidence means. Your argument can be strengthened or weakened at any of these steps.

If your readers can't figure out what problem you are addressing—this does happen—they will doubt your judgment. If they don't think you've worked out an effective approach to the problem, they will think you are missing the point. If they think that your evidence is weak, they will question your rigor. If they don't like the way you've used your evidence, they will question your reasoning. And if you manage to satisfy your readers on all these points, they may still argue that you came to the wrong conclusions.

One way to construct arguments is by linking together:

1. A problem or situation to be remedied
2. A claim or thesis that resolves the problem
3. Background issues that give the particulars of the problem and establish criteria for solving it
4. Evidence that applies the criteria to support the claim or thesis
5. A discussion in which the evidence is weighed and shown to support the claim

This general structure (Figure 4.4) is used repeatedly, quite independently of a writer's field of specialization, profession, or job definition. These elements of argument do not always have to be explicit, although they usually should be. Moreover, the order and extent of their development will depend on the kind of audience you are addressing. The more expert the audience, the more detailed your arguments have to be.

In the short memorandum shown in Figure 4.4, we see these elements in an argumentative sequence. We see a problem that has a commercial and a technical aspect. The problem definition is narrow, but the argument, in this instance, is merely that a formal project should be undertaken to establish the effectiveness of the proposed solution.

Memorandum

TO: M. White, Manager, Support Products Division

FROM: J. Kline, CDT

SUBJECT: Product Developments—A New Scale Inhibitor

DATE: July 22, 1994

Scale Inhibitors—Need for High-Temperature Performance

Condex, Inc. currently markets no effective scale inhibitor for controlling calcium carbonate deposits in multi-stage flash evaporators operating in the temperature range of 240–280°F. We may need this product in order to maintain a strong competitive international standing as production contractors for water distillation plants.

Some informal initial laboratory tests indicate that Condex's DPT-62 might be an excellent potential scale inhibitor that would perform well at temperatures from 230–285°F.

Problem defined both technically and commercially

Stability Requirements in Relation to DPT-62

The main technical problem in developing a new scale inhibitor is thermal stability. Although there are several candidates among the polyacrylates, few of these are stable over temperatures of 240°F. Yet, new flash-evaporator technology will routinely operate at temperatures of 260–280°F within 5 years. Our standard scale inhibitor, H_2SO_4, is too caustic at temperatures higher than 240°F.

A good commercial $CaCO_3$ inhibitor would:

1. Prevent the formation of scale at temperatures between 240–280°F
2. Have thermal, hydrolytic, and oxidative stability at temperatures between 240–280°F
3. Conform to EPA standards for toxicity in water-treatment environments

Lab tests, using a scale adherence simulator (SAS), show that DPT-62:

1. Inhibits $CaCO_3$ scale at temperatures between 220–280°F
2. Has thermal stability at 240–285°F
3. Has hydrolytic stability at 240–275°F
4. Has oxidative stability at temperatures between 240–270°F
5. Is effective in doses that will not exceed EPA standards

Evidence that addresses the problem and supports the conclusion

Recommendations

Laboratory testing, carried out within CDT over the past two months, has established the effectiveness of DPT-62 under most of the conditions set. I recommend:

1. Initiating a full field-test of DPT-62 at the Woods Hole, MA site
2. Building a new laboratory testing unit (SAS) for further scale inhibition studies on magnesium hydroxide
3. Initiating comparative performance tests on the competition's inhibitor, T-XYZ

Conclusion consistent with the aim of the document

Figure 4.4
Argumentative structure in an internal memorandum. This brief recommendation report displays the essential problem identification, followed by the criteria, evidence, and conclusion.

A problem, once divided into so many issues, now demands certain kinds of evidence. Hence, an argument should be considered a system of prose elements, a conceptual plan or prose strategy for making the most of your evidence. In Figure 4.4, the authors have limited the terms of the problem-solving effort to the laboratory checking of thermal stability of the proposed scale inhibitor. They have established criteria for the test, even though in two instances the performance of the $CaCO_3$ inhibitor did not reach the upper limit of the test criterion.

Evidence can be judged only in the context of such criteria, and it may not always perfectly fit the conclusions. But evidence is always interpreted in some larger context of possibilities and needs.

Get Perspective on Your Work

Eventually, your draft will be ready for reviewing. Before your work is in final form, you will want the appraisal of your colleagues and probably your supervisor as well. The time it takes to circulate your document so that others have a chance to read it and make suggestions is your time for a breather. Many writers put their work on the shelf and move on to something else during this period. When they come back to their work to incorporate comments and revise, they have a fresh perspective.

5

Revising for Organization and Style

■

A team of biologists hurries to finish a large proposal for a major funding agency. Pressed for time, team members suspend their laboratory work while they draft their separate sections of the proposal. With 24 hours to go before the deadline, the team meets to read the complete document, revise it, and compile the sections. Unfortunately for team members, their very worthy project may not be competitive. Their aims may be well stated and their plans well outlined, but they've made an all-too-common mistake: They've left too little time for revision.

Revising is part of writing. To achieve the high level of organization and clarity necessary for your writing to succeed, you need to revise your work. Once writing goes out for delivery, the writer no longer controls the message. Your written thoughts are available for unlimited close analysis, "operating" on your behalf in your absence. Good writing can reduce great volumes of descriptive and analytical detail to refined, closely reasoned arguments. The finished document must be clear, accurate, and consistent. You won't be personally present to fill in gaps or correct errors and inconsistencies.

When revising, try to improve the substance, organization, and clarity of your prose. Approach this phase with an open mind. Here are some things to consider:

• *Organization.* Can you reorganize the draft in ways that will make its structure more closely reflect your goals and the needs and interests of your readers?
• *Clarity.* Will adding, deleting, or rewriting material noticeably strengthen its logic, coherence, and flow for readers?
• *Accuracy.* Is the prose detail sufficiently accurate and complete to support your claims?
• *Economy.* Can you condense or eliminate wordy paragraphs or sentences?

Try to gain perspective on your work by setting your outlines and drafts aside for a time. Ask for input from colleagues and supervisors. If you have been immersed in your project, you may overlook obvious gaps in your writing. You tend to read in missing material. Readers' questions and criticisms will let you know whether you're communicating with your audience. Don't be defensive; you don't have to accept every recommendation.

Organization First, Then Style

The distinction between organization and style is a matter of emphasis, because effective organization is good style. Organizational revision encompasses a larger scale of activities. You add, delete, or rearrange content to improve its logic and focus. This process will lead naturally into reworking individual sentences and rephrasing for greater impact on your audience. You start out trying to make the draft reflect more accurately your aim and writing strategy and you work your way into reviewing the draft for clarity, word choice, and economy.

As Figure 5.1 suggests, manuscripts are most effectively revised and prepared in a certain order. As you work on getting your coverage and organization right, try to develop clearer paragraphs and sentences. The manuscript also needs to be edited for grammar, punctuation, and me-

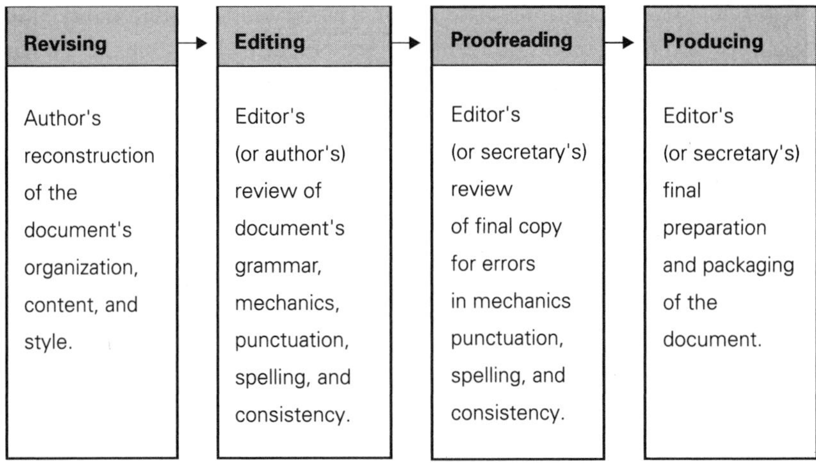
Revising
Author's reconstruction of the document's organization, content, and style.
Editing
Editor's (or author's) review of document's grammar, mechanics, punctuation, spelling, and consistency.
Proofreading
Editor's (or secretary's) review of final copy for errors in mechanics punctuation, spelling, and consistency.
Producing
Editor's (or secretary's) final preparation and packaging of the document.

Figure 5.1
Stages of revising, editing, proofreading, and producing documents. Note that various individuals may play roles at the different stages. For some documents, these stages may not be distinct.

chanics, as well as proofread for various inconsistencies and omissions. Finally, it must be produced for distribution, often to requirements set down by professional societies or agencies. It's good to know how to do all this yourself, even though you may have secretarial or editorial assistance in the later stages of the process.

When revising, first read the draft and circle material that you may want to relocate or delete. Jot notes in the margins, but don't start rearranging or cutting sections until you have reviewed the whole manuscript. Once you have familiarized yourself with the draft, then study your original outline to see if your manuscript meets your expectations.

Here are some questions the reviser should ask:

• *What results do I want from circulating the document?* What claims are you making? What criteria have you set for achieving your claims? What do you want your readers to do after reading the document? Does your document make this response clear?
• *Will my document work for the audience I am trying to reach?* Is the organization appropriate for your readers? Have you thought carefully about material they need in your document to reach their goals? Is the level of your material (including your graphics) appropriate for the audience?

• *Are the aims of my document clear?* Does your introduction contain your document's aim and problem statement? Have you stated the problem clearly? Keep in mind that your problem statement may not become clear until you have worked it over in draft and can see how the criteria and evidence have lined up. Don't elaborate in such detail that the problem becomes obscured by the detail.

If you decide that you want to reorganize your document, consider revising your outline to establish a new plan. Then cut-and-paste sections in the new sequence, using a photocopy or duplicate word-processing file, so that you can retreat to the original version if the rearranged draft doesn't come out right. Remember that once you have cut and rearranged the draft, you will probably have to write new transitions.

Tightening the Organization of Your Draft

Much revision is devoted to reorganizing and developing material. For example, your theoretical section may belong not after your introduction but after the discussion of your experimental results. You may have given an account of your methodology in the results section of your report and want to shift this preliminary material back to its own section. You may want to give a more detailed account of the problem you are addressing. You may have buried important recommendations in back sections, where the managerial reader will have trouble finding them. These and similar problems would require you to rearrange considerable amounts of material.

You often need to develop material that has not been smoothly or logically presented in the draft. Revising is partly a matter of rearranging sections to (1) tailor the detail to specific aims and audiences, (2) improve logical development of the material, (3) add or delete detail for the sake of argument, or (4) highlight important material. Here are some of the common approaches to organizing your draft more effectively.

Reorganizing for Greater Audience Impact

To reorganize, work with an outline, which you can use as a plan for rearranging or deleting. Note, for example, the two outlines for an in-house administrative report shown in Figure 5.2. The authors are recommending office automation and in the first version have approached the question logically, sorting out and discussing the significant factors of

Office Automation at LMN, Inc.: A Two-Year Plan

Logical Order

[1 **Background: problem, aim, criteria**]

2 **Local area networks (LANs)**

2.1 Baseband Systems

2.1.2 Software

2.1.3 Applications and future developments

2.1.4 Cost and availability

2.2 Broadband systems

2.2.1 Information transmission method

2.2.2 Software

2.2.3 Applications and future developments

2.2.4 Cost and availability

3 **Private automatic branch exchanges (PABXs)**

3.1 Information transmission method

3.2 Software

3.2 Applications and future developments

3.4 Cost and availability

4 **Recommended system: PABXs**

4.1 Advantages over LANs

4.1.1 Efficient Software for voice and message transmission

4.1.2 High node capability

4.1.3 Optimal voice and data integration

4.1.4 Adaptability to different vendor hardware

4.1.5 Low installation costs

4.2 Some disadvantages

4.2.1 Data transmission capacity

4.2.2 Distributed data-processing speed

5 **Implementation of a PABX system at LMN, Inc...**

Order of Importance

[1 **Background: problem, aim, criteria**]

2 **PABXs: the recommended system**

2.1 Advantages over LANs

2.1.1 Efficient Software for voice and message transmission

2.1.2 High node capability

2.1.3 Optimal voice and data integration

2.1.4 Adaptability to different vendor hardware

2.1.5 Low installation cost

2.2 Disadvantages

2.2.1 Transmission methods

2.2.2 Distributed data processing speed

3 **Alternative systems: baseband and broadband LANs**

3.1 Transmission methods

3.2 Software

3.3 Most appropriate applications

3.4 Transmission methods

4 **Implementation of a PABX system at LMN, Inc...**

Figure 5.2

Reorganizing content for audience impact. In this example, the material is rearranged from logical order to order of importance as a way of supporting decision making. Note that the shift not only reorders but also condenses the original material.

the technology. Organizing the material by discussing each type of system helps the authors learn the issues but does not meet the needs of the audience, who have to take action.

In the revised organization, the writers reflect a pragmatic order of importance that will appeal to decision makers. To gain acceptance for the recommended system, they first discuss the advantages of private automatic branch exchanges (PABXs) of local area networks (LANs). Having firmly established the reasons for their choice of system, they survey the alternatives. Note that considerable material has been condensed in the second version to deemphasize the systems not recommended. This kind of large-scale revamping and condensation of draft documents may be worked out first in an outline. Once you have the basic structure determined, you can review and revise with other goals in mind.

Making Fragmentary Prose Coherent

Be sure you convert your outline to prose. A frequent problem in draft technical documents is that sections of the draft may still read as though they had been lifted from an outline. The result may be an underdeveloped and poorly focused passage, as shown in the example below:

Underdeveloped writing

3 Experimental procedure

3.1 Apparatus

3.1.1 <u>Chopper wheel</u> (Figure 3):

Diameter—approx 1.2 cm

Material—cadmium, slit width approx. 1 mm

No. of slits—6

Its purpose is to chop the neutron beam into clusters of particles, whose energy distribution will correspond to the Maxwell-Boltzmann distribution

3.1.2 <u>Motor (low-power AC)</u>. Its purpose is to rotate the chopper wheel.

Revised version

3 Experimental procedure

3.1 <u>Apparatus</u>. The main apparatus used for neutron counting is a chopper wheel, shown in Figure 3, that chops the neutron beam into

> clusters of particles. The energy distributions of the clusters should correspond to the Maxwell-Boltzmann distribution. The wheel, as shown, is a cadmium disk, approximately 1.2 cm in diameter, with 6 slits, each approximately 1mm in width. The wheel is rotated by a low-power AC motor.

Your paragraphs should have clear topic sentences. The remaining detail should then be subordinated to the main topic. These measures will help develop the logical flow and coherence of your material.

Expanding the Draft in Order to Clarify Your Arguments

Sometimes draft material needs to be expanded in order to show steps in logic or to provide clarifying detail. Draft sentences may be overloaded with topics that need to be developed more gradually. In the sample below, for example, the prose is compressed and vague. The material needs to be sorted and developed with more care.

> *Underdeveloped writing (2 sentences)*
>
> The current photovoltaic industry remains a high-cost industry, whose production of solar cells is labor-intensive and therefore costly, and this limitation makes investors reluctant to invest because of the small profits involved. According to recent studies (Clifford 1998), this is a problem for manufacturers, who must decide whether to further intensify their labor for possible short-term profits or to seek government support and to automate.

Although the paragraph above is comprehensible, the first sentence attempts to make so many different points that it sacrifices focus and specific detail. Phrases like "high-cost industry" and "small profits involved" are weak generalizations.

> *Revised version (4 sentences)*
>
> The high prices associated with small-scale production of solar cells have prevented the photovoltaic industry from attracting the investment capital needed to automate manufacturing lines. Moreover, large companies in the United States will be unwilling to invest in photovoltaic array production until an annual market of $50 to $100 million is certain (Clifford 1986). In order to meet near-term demand for solar cells, manufacturers may find it cheaper simply

to add workers to the manufacturing line, rather than to invest in new production facilities. This choice, however, is complicated by recent government interest in making large-scale purchases of solar cells in order to stimulate price reductions.

In the revised paragraph, the first sentence is more focused and provides a clearer topic sentence. Rather than including all the information in two rambling sentences, the author now develops the material more gradually in four sentences.

Arranging Material to Frame Discussions More Effectively

The management of detail is a major problem in nearly all scientific and technical prose. Not only are there immense quantities of physical detail, but the detail is also often repetitive. When revising, make use of tables and charts to organize and condense information so that repetitive details don't obscure your discussion. (See Chapter 6.)

Deleting Detail That Does Not Advance the Discussion

Be prepared during revision to cut the many digressions that typically evolve during drafting. Some of these digressions may run to several paragraphs. Does the detail truly support your argument? You might find that some of your information can be deleted or moved to another part of your document.

Highlighting Material That Traces Your Argument

Highlighting and subordinating enhance your organization and should be considered at the revising stage. Several common strategies help improve the focus and clarity of your draft.

- *Repeating major ideas or themes.* Repetition, used sparingly, can help maintain the focus of your work. You can restate a key idea at the beginning of one or two middle sections of the document and again in the conclusion. Such a restatement, as the example below shows, helps to place the new discussion in the context of the paper's main theme.

Paragraph restating a key idea for focus and emphasis

IV. Nematic behavior in other systems

A. Soap solutions. In all of the above systems, we have been dealing with long-range operational ordering (LROO) in liquids

> composed of small, rigid anisotropic molecules. We have seen that LROO (e.g., isotropic-to-nematic phase transition) occurs because at low temperature and high density the decrease in rotational, "mixing" entropy is offset by a lowering of the potential energy and an increase in the translational, "packing" entropy. The details of the isotropic-nematic phase transition depend on.... Thus, it is not surprising that isotropic-nematic phase transitions have recently been observed in systems that have no direct physical connections with the usual liquid crystal circumstances.

• *Using headings and subheadings.* Part of your revising effort should be given to rephrasing headings and subheadings to make them more specific and adding new ones when they will help organize your material. These conventional but often underused aids signal the next topic and help the reader assess the logic and coherence of the material. They announce shifts in the discussion, supply transitions between sections, and together provide an outline. Your system of headings and subheadings should draw directly from your original rough outline.

> *Use of a heading to announce a shift in the discussion*
>
> ... Now that we have established a basis for understanding how a telecommunications system performs its functions, we shall describe how such a system was implemented on the *Voyager*.
>
> Telecommunications Design on the Voyager Spacecraft
>
> The design of any telecommunications system begins with an examination of the mission requirements for telemetry, command, and radio metrics. For *Voyager*, the telemetry requirements ...

• *Using graphics.* The revising stage is a very good time to review your need for graphics. Graphics are the most potent techniques for emphasis. They may be used to highlight important subject matter, condense repetitive prose, and delineate concepts too complex to treat in prose. The simple drawing can clarify complex relationships, emphasize key concepts, and communicate more rapidly than prose. (See Chapter 6.)

• *Drawing attention to parallelism.* Itemizing and enumerating are common techniques for creating emphasis by drawing out the parallelism among concepts. As you revise, consider adding these formatting devices for strategic emphasis. In the example below, itemizations draw attention to a series of criteria for patents. The same principle can be used to list recommendations or highlight other parallel elements.

Use of itemization for emphasis

Patents assign to inventors exclusive commercial rights for inventions provided they meet the following criteria:

- The invention must be novel and not obvious. It cannot be known or deduced directly from prior knowledge.
- The invention must be useful. Knowledge itself cannot be patented.
- The invention must be capable of being made available to others by means of specific directions.

Revising for Style

As you work over your manuscript, you will inevitably find paragraphs, sentences, and words that need stylistic improvement. Good style means economy and clarity. It does not seek strict uniformity in prose but does demand vigorous expression. Clarity governs these choices throughout the document:

- *Paragraphs.* At the paragraph level, clarity refers to unity and coherence, normally achieved by keeping sentences on topic and by linking them together with key words and transitions.
- *Sentences.* Clear sentences generally develop from balanced elements that are concise, specific, and directly worded.
- *Words.* Clarity of word choice means appropriate and accurate word usage. For a detailed stylistic review of the principles of revising paragraphs and sentences, consult our Brief Handbook of Style and Usage at the end of this guide.

From General to Specific

As you rework the organization and then style of your document, you're moving from general to specific parts of your document. Starting with the big picture means that you will clarify your aims, refine your organization, and make necessary changes in your coverage. These alterations in turn will determine many smaller revisions as you analyze your sentences for structure and word choice.

Before you're finished, however, you need to think about illustrations. Graphics, usually labeled as figures or tables, can be as important as prose. They both enhance and summarize information. As you draft any document, you need to think about the visual part of your presentation.

6

Developing Graphics

■

You've spent weeks drafting and revising a report. You've carefully considered your audience and defined your aim. You and your colleagues have read and reread each other's work. Your attention to organization and style has made the final version concise and readable. Looking over the finished report, however, you feel something is missing. Paragraph after paragraph passes without a visual break, without graphic support of your argument. You're still not finished. You need to illustrate your prose better.

Of course, you're unlikely ever to be in such a fix. Engineering and scientific writing so often depend on visual elements that most writers work on the graphics while they draft the text.

Effective graphics condense large amounts of information. They focus attention and reduce data. They are crucial in analysis and argument. Visual representation has been essential to science—from Euclid's

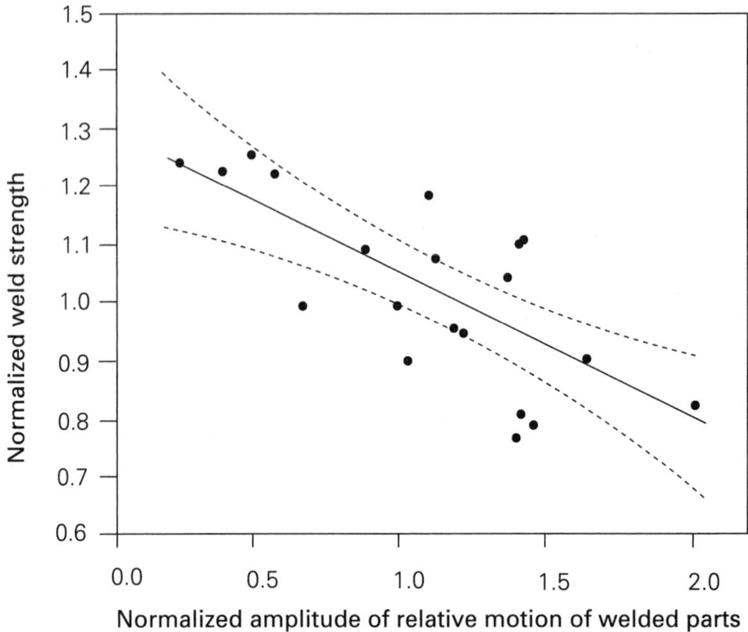

Figure 6.1
This analytical graph correlates two variables: welding motion and welding strength. Each variable is normalized at 1 and thus requires no units of expression. A line showing the trend (linear fit) and two dotted lines showing the confidence limits qualify the data (Source: Jagota and Dawson 1987. Courtesy of ASME.)

geometric forms to Mendeleev's periodic table to Watson and Crick's double helix. An effective graphic promotes thinking and discussion.

Graphics as Analysis and Illustration

A table, graph, or drawing often begins as an analytical tool to help an investigator think about a subject and evolves into an expository strategy to help convince a reader of an argument. The graph in Figure 6.1, for example, summarizes an experimental inquiry into how vibration affects the strength of a welded joint. The scatter chart correlates two sets of data and discovers a trend: The strength of a weld decreases with increased vibration of the parts during welding. This chart, summarizing many hours of experimental work, becomes evidence in a study of welding.

Graphics are also illustrative. The drawing in Figure 6.2a and graph in Figure 6.2b, for example, focus on key elements and dramatically reduce the prose description. These simple visuals quickly identify the variables under consideration. As analysis and illustration graphics serve to do the following:

- *Study numerical data and physical design.* Graphic analysis can help you define and think about your subject.
- *Condense information.* Visuals can summarize information. Use them to concentrate material and to provide overviews of large amounts of data.
- *Improve audience appeal.* Tables, graphs, and diagrams can clarify information for your main audience and attract audiences that would not normally read your work.
- *Focus the argument.* Graphics isolate key items and discard the irrelevant. Use them to focus your discussion on subject matter that's pivotal to your aim.
- *Support the discussion.* Visual evidence often comes closest to demonstrating the phenomenon in question.

Using Graphics to Explore Data

Developing an effective graphic often means transforming raw data into visual patterns that advance the discussion. Simple tabulation such as that shown in Figure 6.3 is normally the first step in putting the data into order. Assume that the table has been prepared by a team using instruments to monitor changing concentrations of particulate matter in the Beaufort Sea, off the northern coast of Alaska. Drafting their report, team members first arrange their data according to specific dates and depths. Such tabulation is the simplest way to organize information. We can see the entire three-month data collection summarized and sorted. Tables are tiny databases that preserve data in accurate numerical form. Yet they also sacrifice point of view. This table does not show trends or reveal specific patterns. Hence it may do little to advance discussion.

Tabulation is the first step in preparing statistical graphics: line graphs, bar charts, histograms, scattergrams, logarithmic charts, and the like.

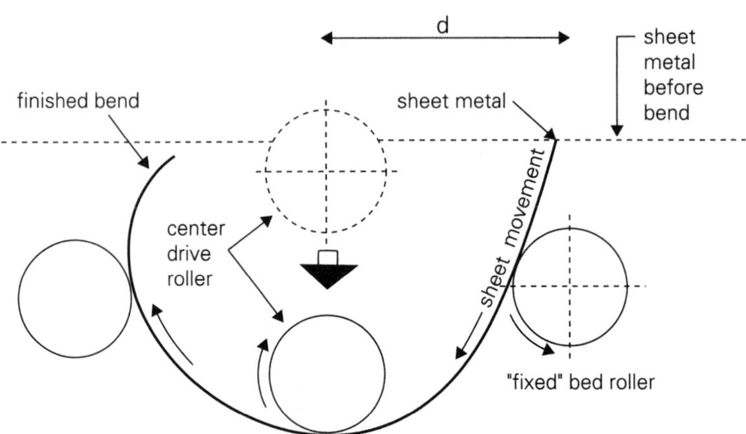

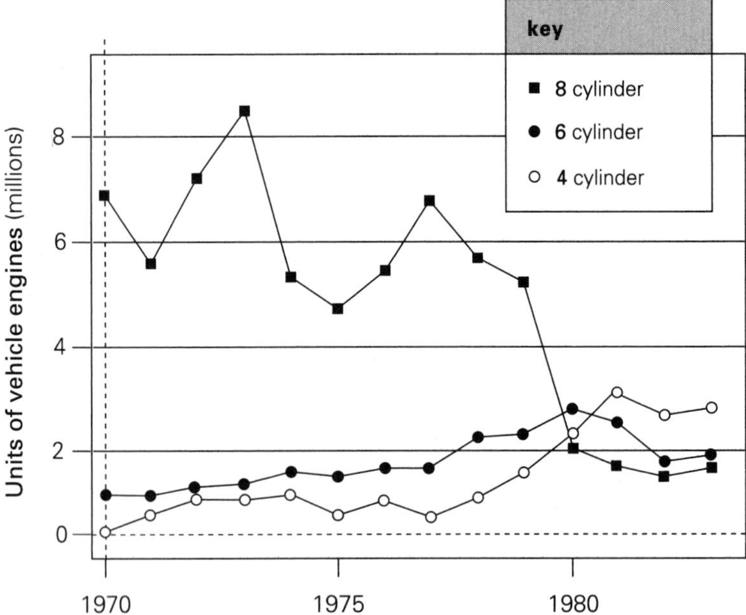

Figure 6.2
Illustrations focus material and sort information. (a) Schematic illustration of a sheet metal bending machine (Source: Hardt et al. 1982. Courtesy of ASME.) (b) Illustration of trends in auto engine manufacture over a 13-year period.

Title → **Table 1.** Concentrations of total particulate matter, particulate calcium, and particulate aluminum in the upper 100 m of the Beaufort Sea.

Columnhead → **Sampling date (1989)**

	Depth (m)	Apr 10	20	30	May 10	20	30	Jun 9	19	29	Jul 9	19
Subhead → Depth; Column → 20												
		Total particulate matter (µg/liter)										
	10	49	180	129	86	45	37	38	61	61	44	60
	25	83	116	72	78	105	19	30	68	46	44	37
	50	132	108	131	77	43	28	32	19	48	34	36
Row →	100	24	20	52	52	28	18	21	25	32	24	26
		Particulate calcium (µg/liter)										
	10	2.3	11.2	5.4	5.6	0.3	0.3	2.2	2.6	5.4	2.4	3.1
Cell (→ 20)	25	3.1	9.1	3.3	3.0	2.4	0.2	1.5	0.8	4.4	2.5	2.5
Rowstub →	50	10.5	3.3	3.1	3.8	0.8	0.2	2.1	1.3	4.3	2.6	2.6
	100	2.5	16.8	1.5	1.7	0.5	0.1	3.3	3.7	3.1	1.2	3.1
		Particulate aluminium (µg/liter)										
	10	0.16	0.34	0.29	0.99	0.31	0.48	0.14	0.18	0.12	0.10	0.14
	25	0.12	0.27	0.21	0.88	0.50	0.19	0.13	0.44	0.10	0.13	0.10
	50	0.19	0.82	0.17	0.17	0.18	0.10	0.93	0.07	0.05	0.05	0.09
	100	0.08	0.21	0.04	0.06	0.09	0.17	0.62	0.12	0.60	0.92	0.08

Source line → **Source: Beulher and Jacyna (1992)**

Figure 6.3
Tables are the simplest visual format and preserve the original data. Each cell represents a full sentence. Tables do not, however, convey visual patterns and may obscure significant events or trends.

These forms, whether developed from a database and computer graphics package such as Excel (Microsoft Office), SAS (⟨www.sas.com⟩), or Matlab (⟨www.mathworks.com⟩) or even roughed out on a pad of grid paper help you examine your data for important trends. They are thinking aids. Depending on the point you're making, you could create a variety of graphs out of the data in Figure 6.3. Two of many possibilities are shown in Figure 6.4a, b. The simple line graph compares two trends in the data; the semilogarithmic chart compares rates of change for two series of values that differ greatly in magnitude.

Choosing the Type of Graphic

Modern data graphics libraries offer an immense variety of options for analysis and presentation; familiarize yourself with those that are used in

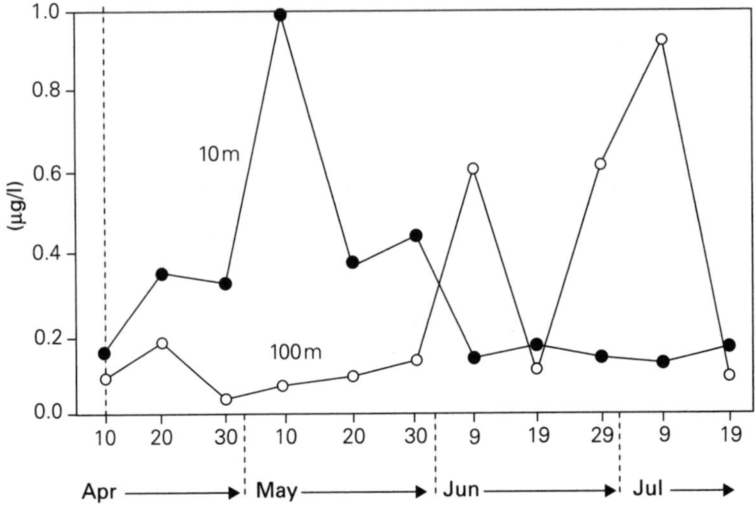

(a)

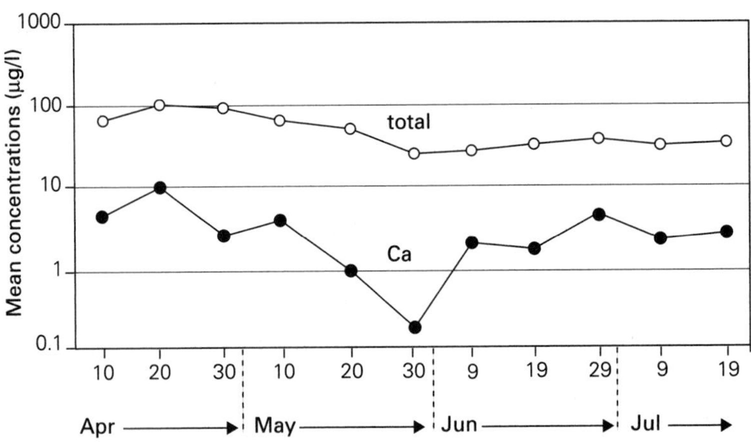

(b)

Figure 6.4
Graphics can manipulate data. Many different graphics may be prepared from the tabular data in Figure 6.3, depending on the writer's purpose and audience. The line chart (a) focuses on the comparative flux of one element at different ocean depths. The semilogarithmic chart (b) plots the mean values of two data series of very different magnitudes.

your field. A simple, widely available package like Excel enables you to visualize data in a variety of line graphs, bar charts, surface charts, scattergrams, logarithmic charts, pie charts, and others. The more elaborate suites of programs in Matlab and SAS apply to a variety of specialized fields in the sciences, applied sciences, and social sciences.

Choose a graphic design that supports your argument and works effectively with your data. The choices range from simple line graphs that track a set of values for a given item to more complex series that plot correlations of two interrelated variables. Some of the more common kinds of graphic visualization include:

• *Items with Different Values.* Tracing the different values for one or more items is one of the most common kinds of graphic display (Figure 6.5). These values may be plotted as bars on a bar or column chart or data points on a line graph. An example might be a series of values for costs of a specific service such as electrical power in different parts of the country. Bar charts are the most straightforward way to visualize this kind of discontinuous data. Sometimes called column charts if they are aligned vertically, bar charts are effective for showing a series of discrete values. Pictographs, which are bar charts whose bars are made up of symbols, are often effective with general audiences. You might use a line graph, however, if you are showing a *trend* in the changing values for an item: the changing fructose content in a species of plant, for example, as it is sampled at latitudes progressively closer to the equator.

• *Time Series.* Time series, probably the most widely used of all data graphs, plot a changing value for one or more items in relation to some unit of time (see Figure 6.4). This kind of *continuous* data, marked by uninterrupted extension in time, is best represented in a line graph that connects the data points. Data appropriate for time series might be the size (area) of a bacterial culture per one-hour period under a defined set of conditions. Other, less continuous kinds of time-related data, such as the sales volume for an item in liters, kilograms, or dollars over several months, may be represented better in a bar chart. It is possible, of course, to connect the ends of the bars (at their midpoints) in a bar chart with a line to emphasize a trend in discontinuous data.

• *Percentages.* Line graphs, bar charts, and pie charts can all express percentages (Figure 6.6). Percentages may be parts of a whole, or they may be percentiles, which express more subtle distinctions. For example, you could represent the percentage of ozone in air samples at 100-meter increments between 5,000 and 10,000 meters, or you could graph the percentage of a specific population or percentile of students scoring A, B,

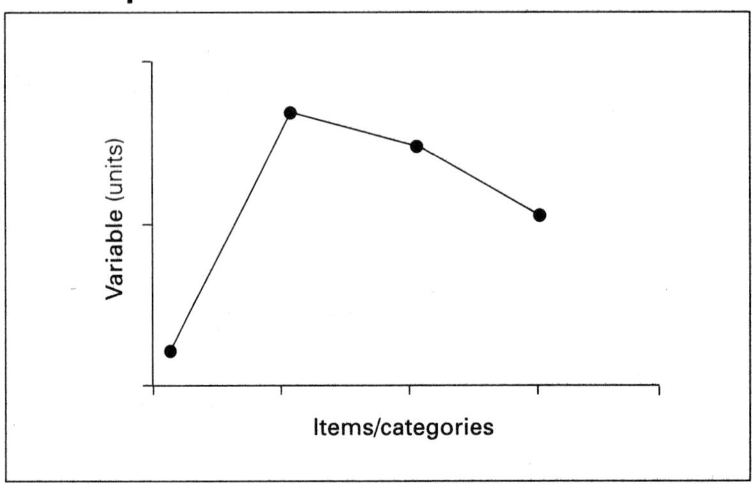

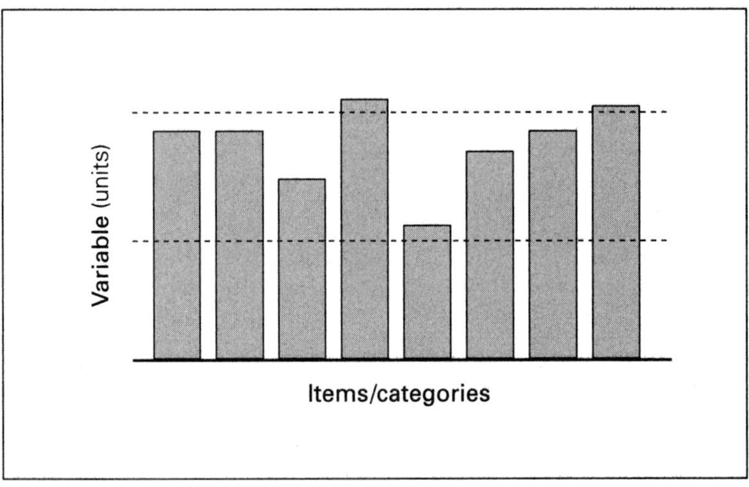

Figure 6.5
Typical graphs of items with different values.

Line Graph

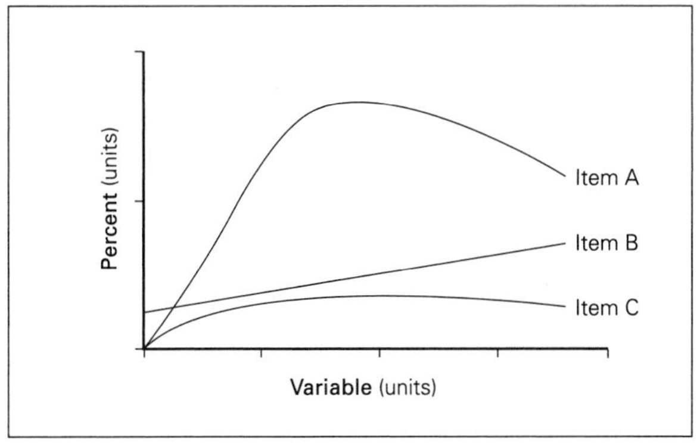

Bar Chart

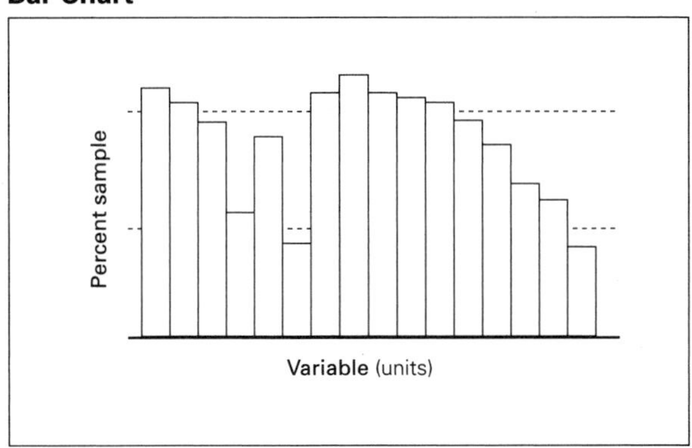

Pie Chart

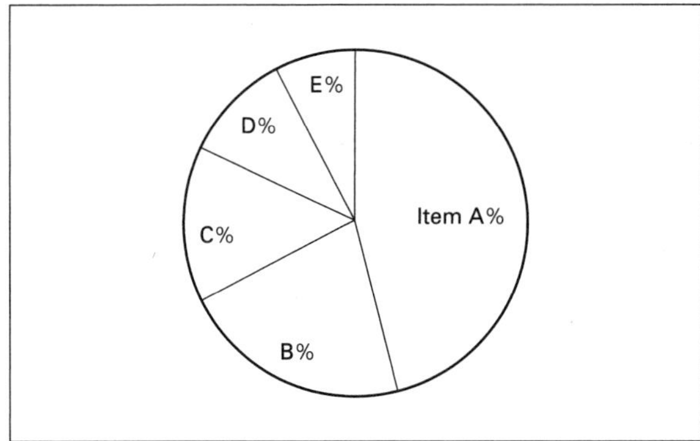

Figure 6.6
Typical graphs of percentages.

C, D, and E points on a given exam. If the percentage readings are taken at very frequent intervals, then a line graph is a good choice. For larger intervals, the bar chart is more appropriate. For a simple percentage breakdown of a given sample, the pie chart or the 100 percent bar chart are the most effective.

• *Comparison.* In plotting comparisons (Figure 6.7), your challenge is to design a graph that displays a series of trends simply and clearly. Pie charts are not as effective for comparing values as the line graph or bar chart. Multiple data fields are very effective when the scales of the data trends are changing or when there are too many trends to fit into a single data field.

• *Correlation.* Correlations demonstrate or suggest the mutual influence (covariance) of two variables (Figure 6.8a). The variables *A* and *B* are normally two different items or two aspects of a single item. The correlation in Figure 6.1, for example, relates the strength of a weld to the motion of the welded parts during welding. Electrical output is a function of resistance, and bacterial growth is a function of ambient temperature. Correlation represents these linkages.

Each variable depends on the other, or both variables depend on yet another phenomenon. The covariance may be positive or negative, where an increase in *A* shows a proportional increase or decrease in *B*. The independent variable is plotted on the horizontal scale for a line graph or scatter chart. The line graph and scatter chart are most effective for displaying correlation because each data point is an expression of the two variables.

Be careful not to confuse correlation with comparison. Correlated variables are supposed to be interdependent, although proof of their interdependence is often the key issue in the graphic display.

• *Ratios and Rates of Change.* The semilogarithmic scale (Figure 6.8b) is an effective way of plotting data for an item whose values vary greatly. For example, if the series of data points jumps from the 10^1 range to the 10^3 range, the trend cannot be seen clearly on a standard scale because a scale designed for the 10^3 range would not show variations at the 10^1 magnitude. With the semilogarithmic scale, which shows rate of change, different magnitudes may be plotted in detail and with clarity. An increasing slope shows increasing rate of change in relation to prior values.

• *Frequency Distributions.* Distribution graphics illustrate the spread of a population, as in the bell curve distribution demonstrating probability (Figure 6.8c). In constructing frequency graphs, you select an incremental unit (the independent variable) that will best reveal the shape

Line Graph

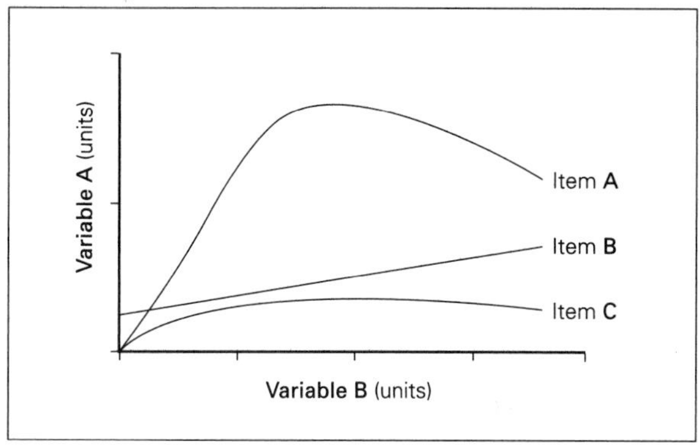

Bar/column Chart

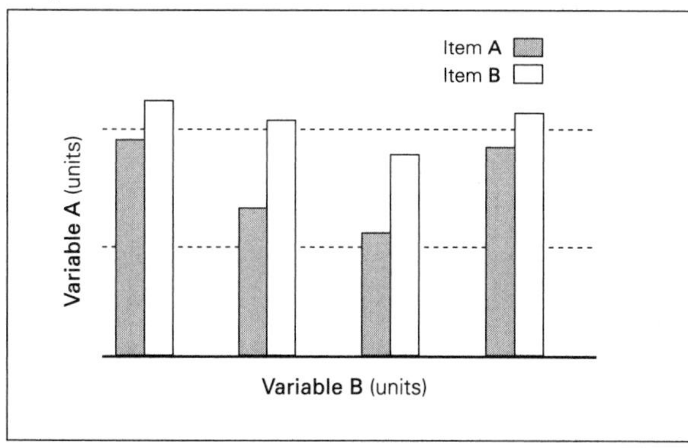

Multiple Data Fields

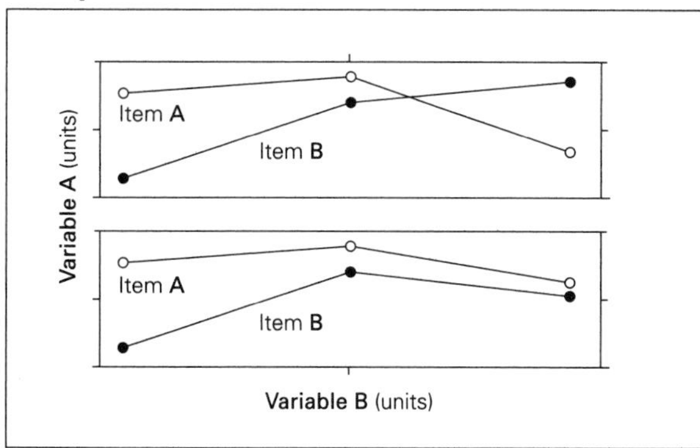

Figure 6.7
Some typical ways of graphing comparisons.

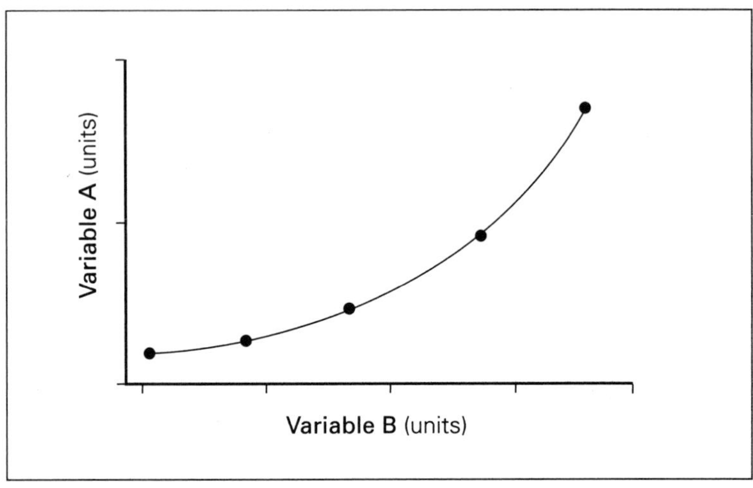

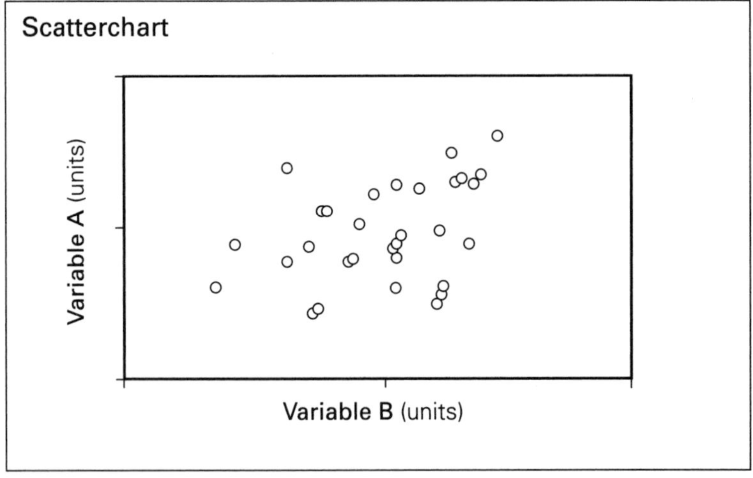

(a)

Figure 6.8
Some common varieties of analytical graphs: (a) correlation of two variables, (b) rate of change, (c) frequency, and (d) net differences.

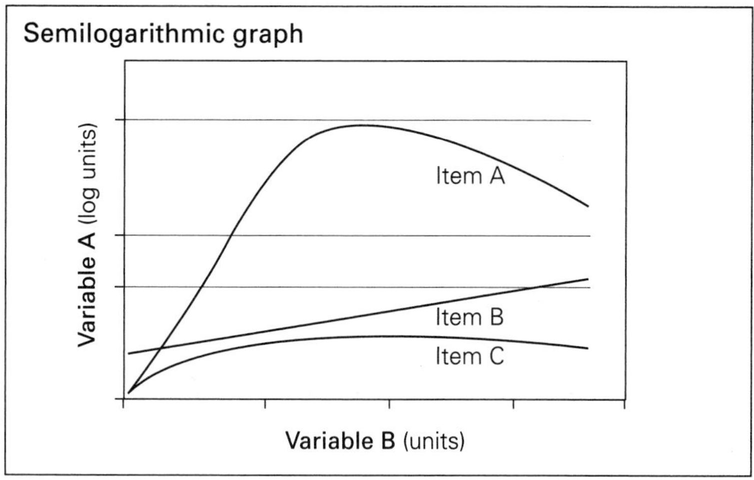

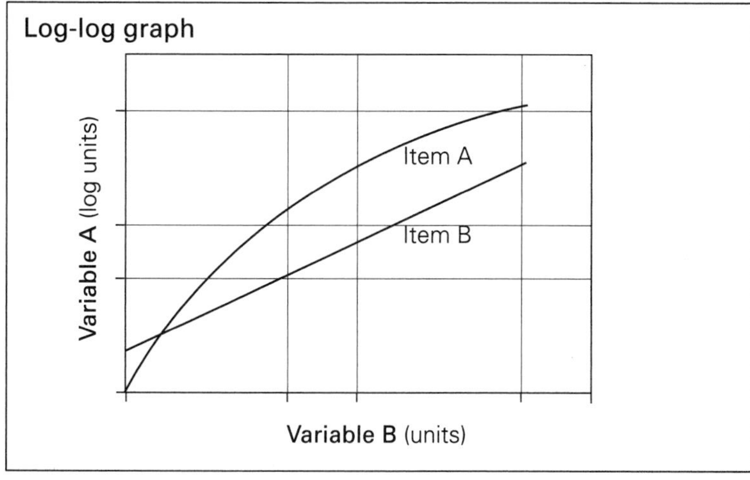

(b)

Figure 6.8 (continued)

Frequency

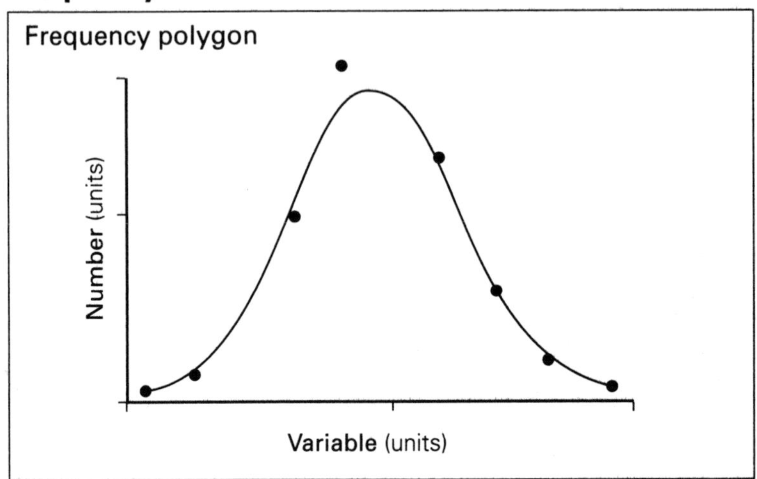

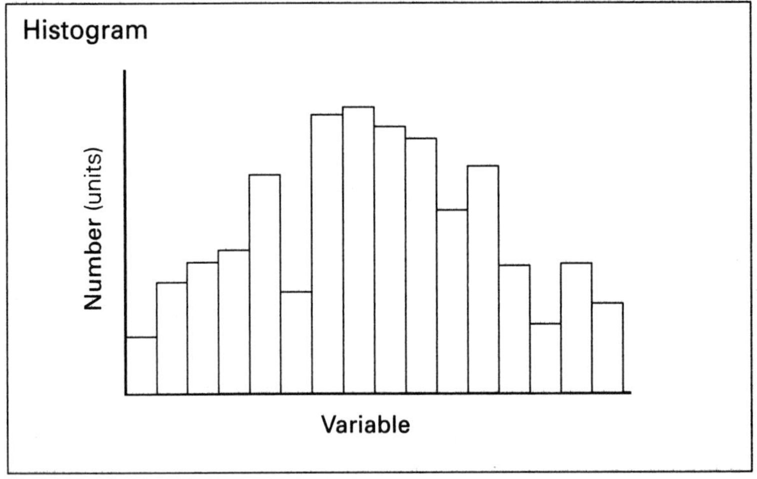

(c)

Figure 6.8 (continued)

Net Differences

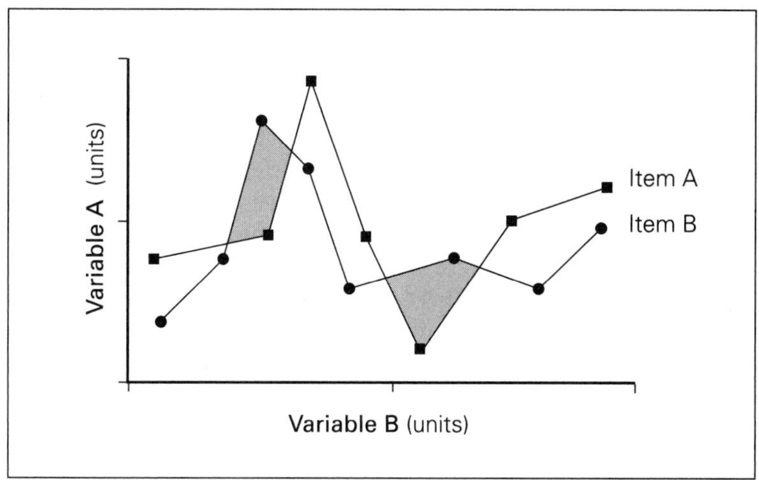

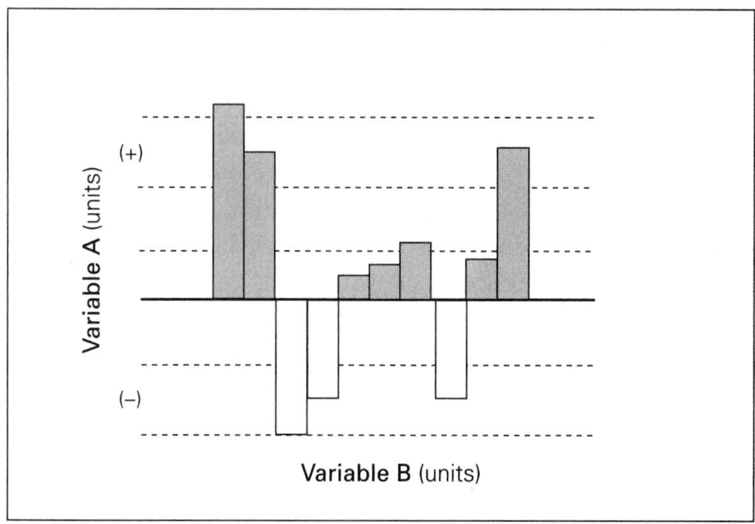

(d)

Figure 6.8 (continued)

of the sample population and then plot the number or percentage of items in the total population that belong in each increment.

Some variables, such as electric current or heat, are continuous, whereas others, such as population, are discontinuous. Continuous variables are best represented with a line graph, a frequency polygon (with data points connected by lines), or a smooth curve. Discontinuous variables are best presented in histograms or step charts (Figure 6.8c).

Note that in your frequency graphs and charts, the plotting of data will depend on data collection. Data may be plotted for specific values (e.g., 0° or 5 °C, Day 5 or 10), or data may be plotted for intervals (e.g., 1–4 or 5–9 °C, Days 5–9 or 10–14). For specific values, the data point appears directly above the value on the horizontal scale. For intervals, the data point appears above the midpoint of the interval on the horizontal scale. For a step chart or histogram, the interval (i.e., 5–9 days) is the width of the column.

• *Net Differences.* Net difference charts use both positive and negative values to show fluctuation and resulting differences (Figure 6.8d). Deviation bar charts illustrate the percent deviation from a given value, such as a mean value.

Preparing Tables, Graphs, and Drawings

Tables

Tables represent portions of your data that are important enough to merit inclusion in your presentation. They may conveniently be prepared using spreadsheets, although you may wish to override some elements of the template in order to simplify the table. A table like the one in Figure 6.3, prepared in Excel, gives us a clear, detailed, and numerically accurate snapshot of hundreds of hours of work.

Because there is so little supporting prose to make sense of a table, it is important to design a simple structure, using as few lines as are necessary. Give special attention to creating a concise title and descriptive headings (i.e., boxhead or stubhead). Place the title above the table. After or under each heading, specify the units in parentheses (see Figure 6.3).

Arrange the independent variable of the table along the horizontal axis and the dependent variable along the vertical axis. Number each table consecutively as Table 1, 2, 3 (larger documents with chapters use Table 1-1, 1-2, 1-3. or 1.1, 1.2, 1.3, and so on). Align the values in the vertical columns along the decimal. If you are missing a value for a cell, leave the space blank. Be sure to identify and discuss the table in your main text.

Graphs

Designing and preparing graphics takes time, even with the powerful assistance of spreadsheets. The table and charts shown in Figures 6.3 and 6.4 will take a skilled user of Excel a couple of hours to prepare. Crude draft graphics should suffice until you're ready to commit the time needed to complete a finished version. Preparing a graph involves a series of steps:

1. *Select a design that demonstrates the point you want to make.* Early in the project, read over your notes, outlines, or drafts and think carefully about which key ideas you want to support with your graphics. Note places as you write your draft manuscript where graphics should be used to support an argument or illustrate a mechanism or process. Study other documents in your subject area to get ideas for effective approaches to displaying data. Using your database, you may want to experiment with various graphic templates to see which ones work well with the data. Keep in mind that developing finished graphics is time consuming, space is limited, and reproducing graphics is expensive.

2. *Choose your scale and plot your data.* When you have decided on a graphic design—items with values, time series, percentages, comparisons, correlations, rates of change, frequency distributions, net differences, or others—work out a scale that will effectively reveal the trend you want to show. Plot the data.

• *Line graphs.* Place the independent variable on the horizontal x-axis and the dependent variable on the vertical y-axis. The intersection of the x- and y-axes is zero. If you start either axis at a value greater than zero (i.e., suppress the 0), indicate this with a broken scale line, as shown in Figure 6.9a. Design a scale that doesn't exaggerate curves—see Figure 6.9d. Keep the data field clear of material other than the data itself, and make the curve lines more prominent than other lines of the graph. These data plotting principles also apply to scattercharts, logarithmic charts, and frequency polygons.

• *Bar charts.* If the bar chart is aligned vertically (i.e., column chart), the independent variable goes on vertical axis; if the chart is aligned horizontally, the independent variable goes on the horizontal axis. If you start plotting the values for your dependent variable at a value greater than zero, indicate this with a scale break, as shown in Figure 6.9d. Avoid using distracting hatching to fill in the bars. Use alternating shades

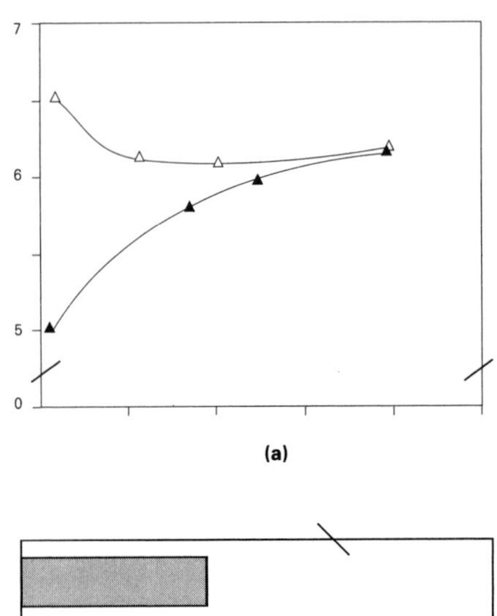

(a)

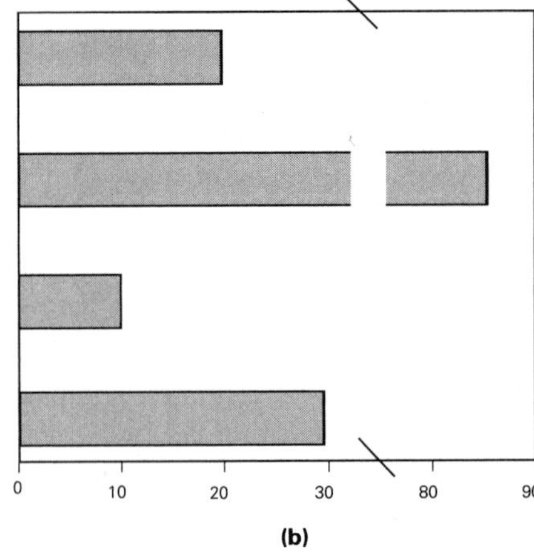

(b)

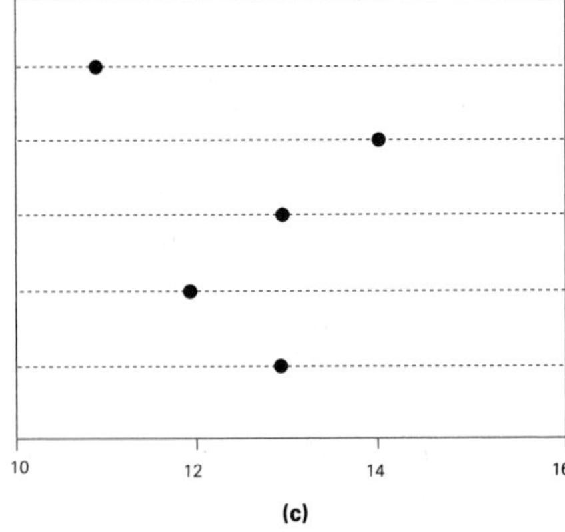

(c)

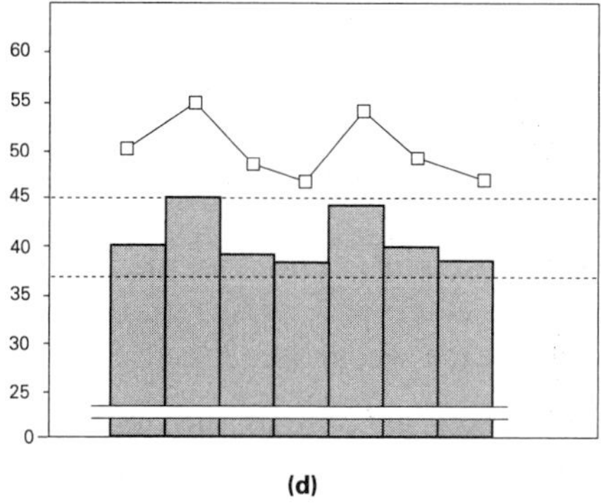

(d)

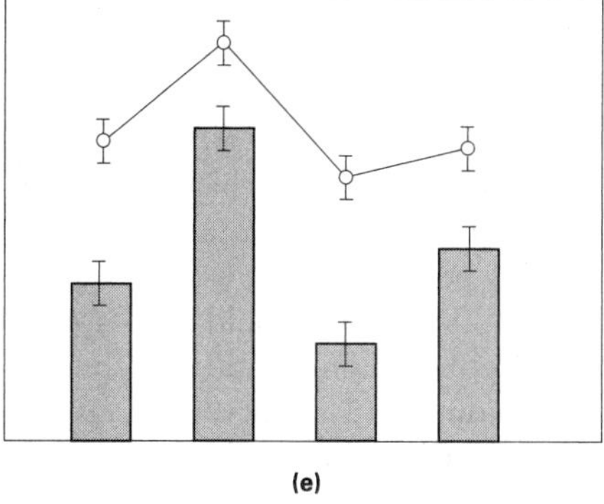

(e)

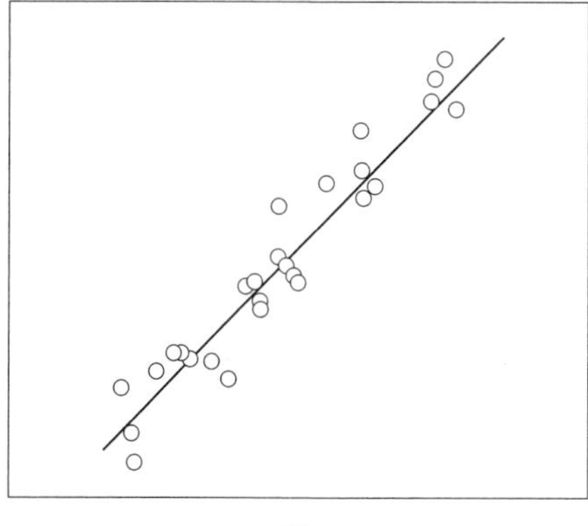

(f)

Figure 6.9
The scale and design of a graph can affect accuracy. The horizontal and vertical scales for line graphs should begin at zero. A suppressed zero (a) or scale break (b) can be used to eliminate a large blank space in the data field. A continuous horizontal scale line across a dot chart (c) eliminates the need for a zero baseline. Scales should clearly reveal variation without exaggeration (d). Use range bars (e) to show standard deviation, standard error, or confidence intervals—be sure to identify which. Theoretical curves and lines with mathematical formulas (f) should be plotted along with empirical data for comparison.

of black or other colors. Dot charts are similar to bar charts, but use a small circle or dot rather than a bar to plot the value—see, for example, Figure 6.9c.

• *Pie charts.* When preparing pie charts, plot sections clockwise, beginning with the largest section at 12 o'clock and adding successively smaller ones. Identify the percentages in each wedge (see Figure 6.6).

3. *Label the graph.* Many otherwise excellent visuals turn out to be useless because they are incompletely or ambiguously labeled. All graph labels start with a figure number, followed by a concise title that clearly describes the trend or content being displayed. In the final document, the figure title appears below the data field. Titles may be followed by one or more sentences, sometimes called the *caption*, that further describe the content of the graph. Number each figure consecutively as Figure 1, 2, 3 (for larger documents with chapters, number figures by chapter as Figure 1-1, 1-2, 1-3 or Figure 1.1, 1.2, 1.3, and so on). If you are using data from a copyrighted source, get written permission to use it and give credit to your source at the end of the caption. You should also label the variables on the x and y scales of the graph along with their units of measurement. If your graph includes comparative curves or multiple bars, label each one directly or in legends that identify the bar or curve according to the shapes of its data points or the shades of its bars (see Figure 6.7).

Drawings and Other Illustrations

Line drawings and other artwork may be prepared in several ways. You may photograph items for reproduction, optically scan printed illustrations and photographs into files for submission with your document, or prepare your own illustrations with the help of a drawing or computer-aided design (CAD) program. Line art, such as the apparatus shown in Figure 6.10, can be prepared on a computer drawing program in a couple of hours. In all these instances, it is important to find out from your editor the requirements for submission so that you don't waste time in formats that the publisher doesn't accept. Always select or prepare materials that contribute to your discussion; resist the temptation to clutter your document with costly distractions.

Give each illustration a figure number and a short descriptive title. Add a succinct explanatory caption if it will help the reader situate the illustration. If you use material from another source, obtain written permission from the copyright owner and identify your source at the end of your caption. Be sure to refer to your illustration in the text.

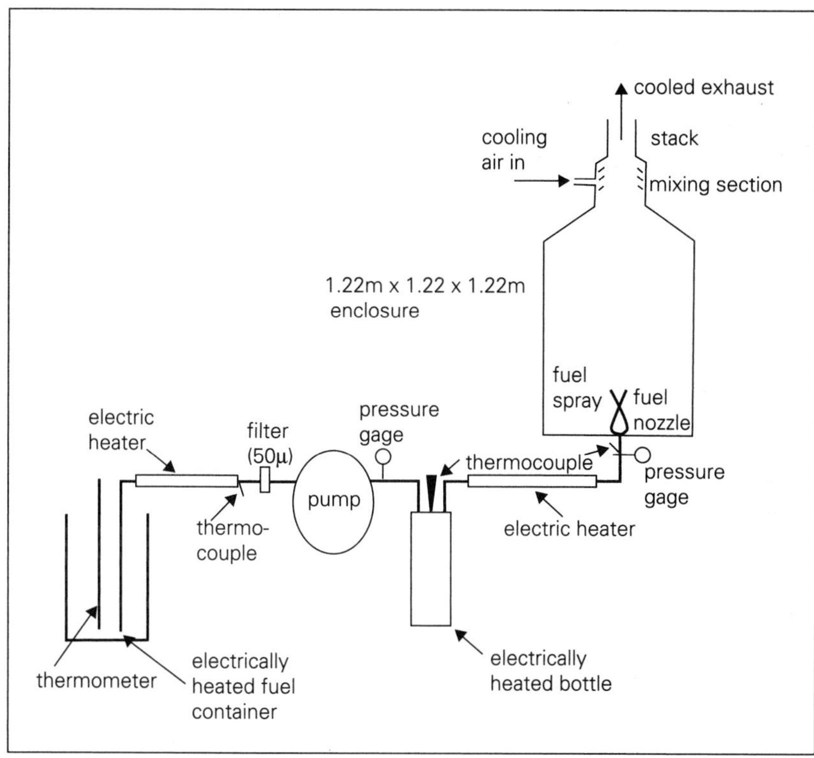

Figure 6.10
A typical schematic demonstrating a process mechanism with flow moving from left to right. The apparatus is an experimental setup for atomizing No. 6 oil. (Source: Kwack et al. 1992. Courtesy of ASME.)

Revising and Producing Graphics

It takes time to perfect a graphic. Some element always needs tweaking before you can be sure that the graphic will be clear to your target audience. Here are some final things to check before you submit for publication.

Are Your Curves and Scales Accurate?

The accuracy of data is often an issue. Authors face two special concerns: the technical accuracy of the data and the presentational accuracy of the visual. Check over your data to be sure you have derived it effectively and review your visualization of the data to be sure you are presenting it with a minimum of distortion. Some of these accuracy issues are illustrated in Figure 6.9.

Are the Graphics Adapted to Your Readers?

Use visuals that work for your audience. Readers vary greatly in their familiarity with various forms. A specialist in a scientific or technical field normally expects a range of interpretive analytical graphics—scattergrams, histograms, logarithmic charts, detailed design drawings, function plots, and the like. The drawing shown in figure 6.10 is a typical no-frills illustration of an apparatus addressed to a technical audience. By contrast, administrators often prefer much-simplified graphics that address their managerial concerns for understanding a situation and allocating resources. Simple tables, line graphs, bar charts, pie charts, and diagrams, produced by spreadsheet programs, are appropriate for such decision making.

Are Your Graphics Consistent with Your Data?

Not all graphics are equal to a given task. You may be displaying the right information without effectively making the point. Be prepared to discard or redesign a graphic if it is not doing the job. We see an example of this problem in Figure 6.11a, which attempts to use pie charts comparatively. The author is trying to argue that aluminum will continue to hold its own with composites (CFCs) and Titanium as a material used to build military and civil aircraft parts. Yet the four pie charts do not facilitate a reader's recognizing either the general trends or the specific instances of aluminum use. The eye has trouble distinguishing the comparative sizes of pie slices in two different pies (not to mention 4). A set of segmented 100 percent bar charts (Figure 6.11b) does a better job of comparing.

Are Your Graphics Used Effectively in the Text?

A graphic not only needs a clear descriptive title, caption and labels; it also needs to be discussed in the text. As you work on improving the clarity of your titles and labels, look for ways of getting more out of your graphic in your textual discussion. Be sure to use the same terms and units in your graphic and your discussion in the text. Tell the reader what the graphic means. Explain how it supports the main discussion.

Add parenthetical information (conditions, limitations) not identified in the graphic's caption.

Is Your Graphic Cleaned Up for Submission?

Because fine details can make or break a graphic, be doubly vigilant as you prepare the final version. Cryptic and incomplete labeling are among the most common sources of ineffective graphics. Emphasize data, not graphic apparatus. Be wary of the fancy background patterns available on the many spreadsheet routines, for apparatus can easily overwhelm the substance of a graphic. Be sure, however, that your apparatus is complete and visible.

For example, a crude graph, such as the line graph of Figure 6.12, will have many minor problems that need attention as you prepare the final manuscript. The example is a rough semilogarithmic line graph developed from an experimental series. The graph is numerically accurate but needs cleaning up and resizing for final production. If the graph were reduced to 60 percent of the original, as many graphics routinely are in final production, the information would disappear. An improved version is shown in Figure 6.13.

Consider the following (see Figure 6.12) as you prepare the final version:

1. *Title and caption placement and submission.* Titles are treated as part of the text, not part of the figure. If you are submitting your figure for outside publication, include the figure number and title in brackets where you think it should appear in the text. Then identify the figure number in the file of the floppy disk or tape the number on the back of the completed figure. When submitting your manuscript, include a separate "List of Figures and Tables" in a floppy file that gives the full number, title, and caption for each of your visuals. (List tables and figures separately.) If you are producing your own documents, titles appear below figures and above tables.
2. *Symbols.* Check to make sure that you are using the standard units and symbols of your discipline. The symbol for "Hertz" is "Hz" (not "HZ").
3. *Title and caption wording.* Wording should be unambiguous. Place modifiers next to words they modify. The modifier "0.25 uF" applies to capacitors and should be better placed.
4. *Data points.* Be sure that data points are large enough not to be obscured when reduced. Squares, circles, and triangles are more effective

(a) Ineffective Comparison Using Pie Charts

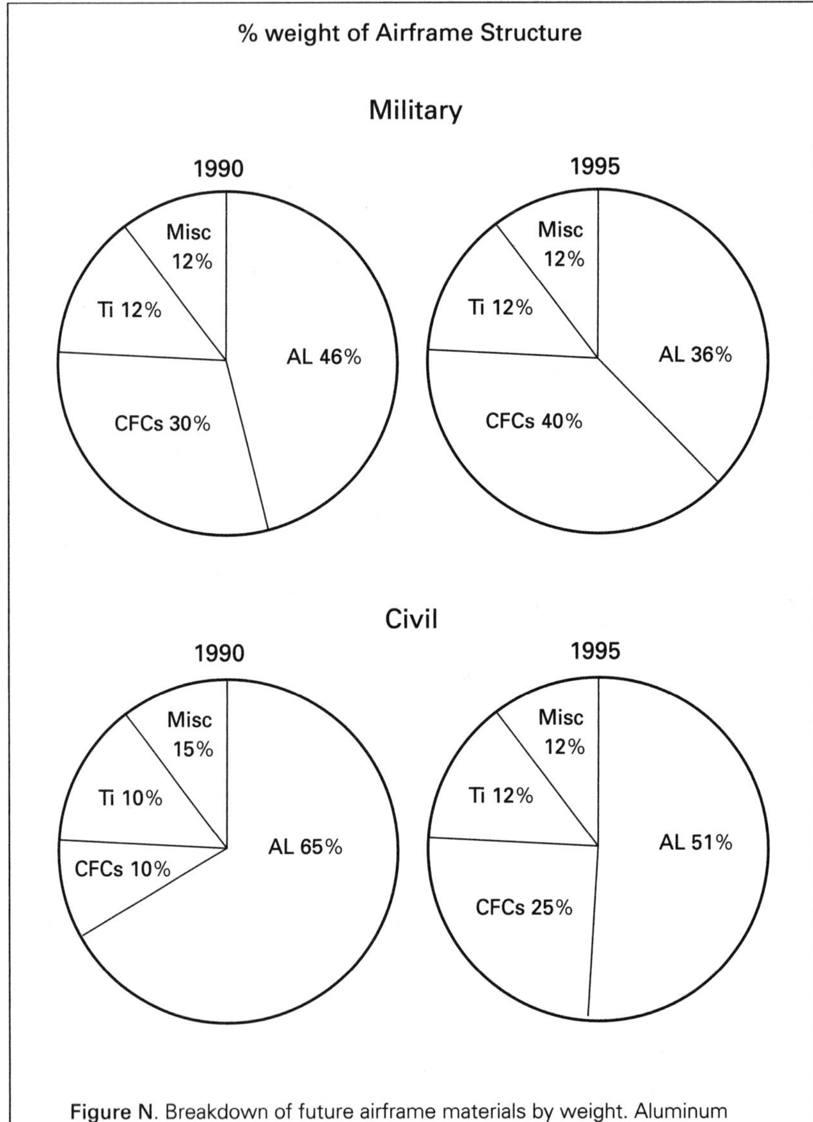

Figure N. Breakdown of future airframe materials by weight. Aluminum industry experts concede that aluminum is going to lose to other materials, but do not believe it will be the dramatic loss commonly projected. An existing technology and cost effectiveness may give aluminum the edge.

(b) Improved Comparison Using 100% Bar Charts

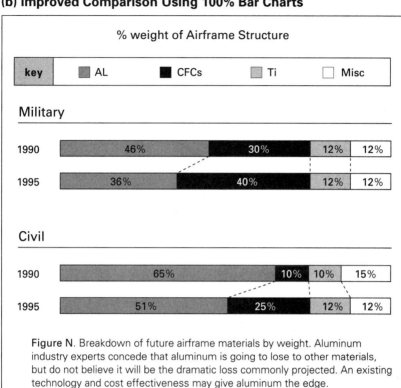

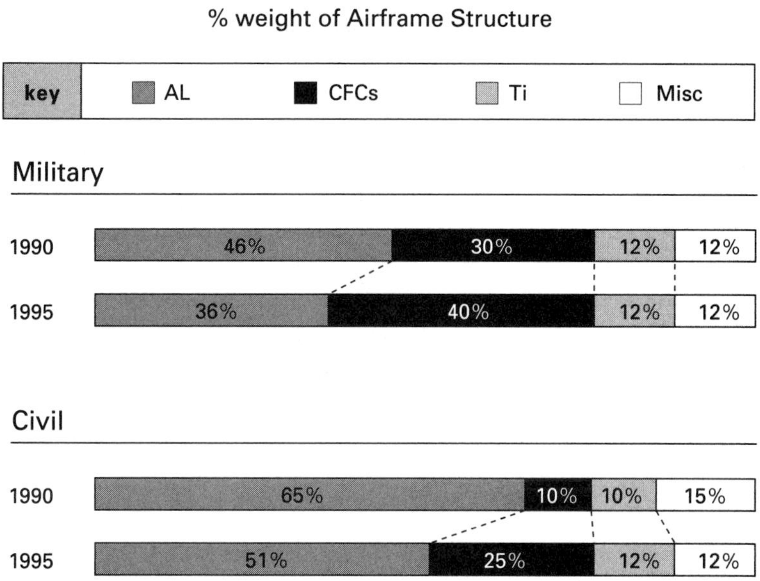

% weight of Airframe Structure

key AL CFCs Ti Misc

Military

	AL	CFCs	Ti	Misc
1990	46%	30%	12%	12%
1995	36%	40%	12%	12%

Civil

	AL	CFCs	Ti	Misc
1990	65%	10%	10%	15%
1995	51%	25%	12%	12%

Figure N. Breakdown of future airframe materials by weight. Aluminum industry experts concede that aluminum is going to lose to other materials, but do not believe it will be the dramatic loss commonly projected. An existing technology and cost effectiveness may give aluminum the edge.

Figure 6.11
Make the graphic support the argument. In (a) the author attempts to use four pie charts to compare component values of four wholes. The comparisons are hard to follow. In (b), two segmented bars make a clearer comparison of the same components. (Adapted from *Aviation Week and Space Technology*, 1984.)

than different styles of lines in distinguishing among curves. Emphasize actual data points and make lines between them less prominent. If the curve itself is a calculation, however, make it the most prominent line on the graph.

5. *Labels.* Do not place complicated legends in the data field. Whenever possible, just label parts of the figure (e.g., curves). Avoid extra comments that clutter your data field.

6. *Scale numbering.* Simplify scale units by placing their multiples (e.g., 1000s or 10^3), just below the scale or in the label.

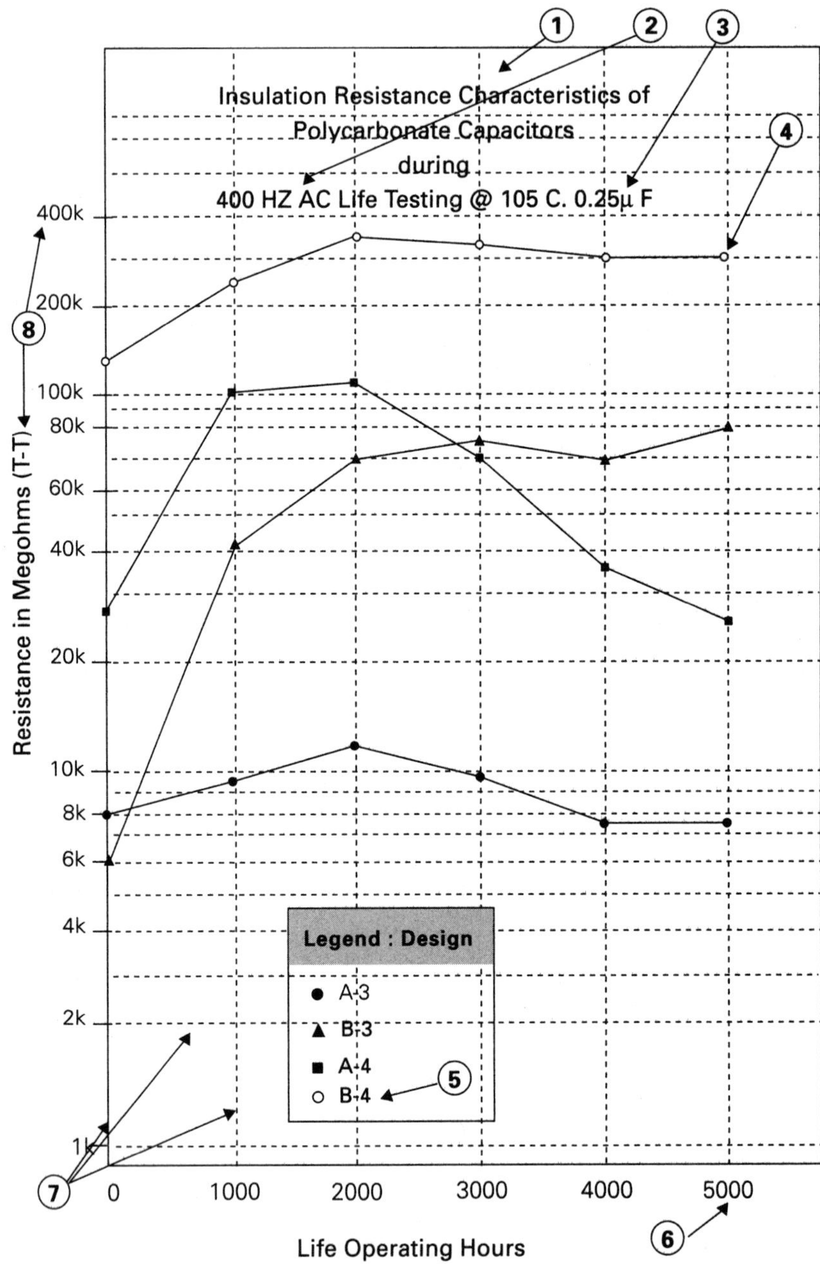

Figure 6.12
A draft graph. Although the information is accurately plotted, the graph is poorly sized for reduction, the labeling is unclear, the data region is cluttered, and the data points are too small for later reduction in publication.

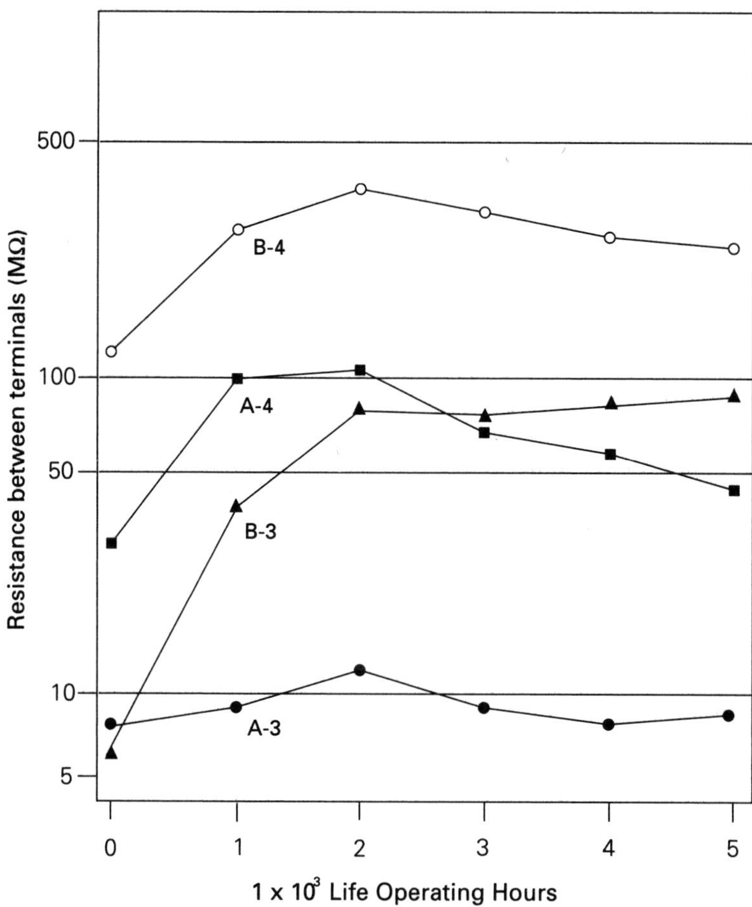

Figure N. Insulation resistance characteristics of 0.25μF polycarbonate capacitors (400-Hz AC life testing at 105°C)

Figure 6.13
This revised version of Figure 6.12 has improved the clarity of the information. The original legend and labels have been reworded.

7. *Boxes, tick marks, and grids.* Locating aids help orient the reader, but use them with restraint. Keep them thin and unobtrusive. If you use a box around your graphic, place tick marks outside the box, where they do not interfere with nearby data.

8. *Nonstandard abbreviations in labels and legends.* Avoid using cryptic abbreviations. When abbreviating, include enough letters to ensure clarity. For example, the "T", which stands for "terminal" in Figure 6.12, should be spelled out in the legend so as not to be read as "temperature".

The Many Uses of Graphics

Graphics are often the most effective elements of documents—proposals, reports, articles, and entire books—where they arrange, summarize, and highlight data. Prepare them well, and they will amply reward you for your time. Plan a graphics strategy at the early outline stage of your document. Simplified versions of your document graphics will also be essential to any oral presentations your give on your topic (see Chapter 15). Live audiences often focus attention best on key points when they are presented in visual aids, whether in overhead transparencies or PowerPoint. A diagram often readily explains what words obscure.

7

Design of Page and Screen

Developing Document Standards
Prepare a Style Guide
Think Visually
Design Individual Pages
Specify Design Elements
Consider Options for Illustrations
Include Navigation Aids
Consider Binding Options
Electronic Document Templates
Issues in Screen Design
The Future of Document Standards

■

For almost two years you've been drafting documentation for a start-up company. You've worked with all departments at each stage of product development. Your efforts have paid off. Now, as the company moves from development to marketing, you've been made the publication manager, and you've just hired two new staff writers. Your first task is to instruct them about specifications for the company's documentation, most of which are stored only in your memory. You need quickly to develop document standards so that all work from your department is consistent. To complicate matters, you want documents to be available both in hard-copy and electronic form.

Technical documents are rarely built of words alone. In addition to sentences and paragraphs, they are likely to contain a mix of other elements: headings and subheadings, headers and footers, tables of contents, indexes, appendixes, and figures and tables. This mix creates design and presentation challenges for both hardcopy and electronic information.

A high-quality document is consistent in the way elements are treated. Page numbers appear in the same location on every page. The same font is used for all first-level headings. All bulleted lists are indented the same number of spaces. Abbreviations, acronyms, and plural formations are standardized. Equations are either numbered or not numbered, but the practice is unchanged throughout all documents in a series.

The widespread use of computers to communicate scientific and technical information has created more interest than ever in standards for format and style. Required page formats can be stored as electronic templates. Writer's choices over matters of language, structure, and design elements can be limited so that consistency is achieved across an entire document or set of documents.

Developing Document Standards

Writers in government, corporate, and academic settings are rarely the first people in their organization to write memos or letters, to prepare a set of presentation slides, or to bind progress reports. Many work settings already have standard formats for memos and letters, preprinted covers for reports, templates for transparencies and slides, and "house style" for settling questions about the treatment of oversized illustrations.

In some work settings, document standards are transmitted informally: Authors ask colleagues who have already prepared similar documents, or they examine models of earlier work. In other settings, standards are transmitted through a written style guide. The style guide may have been especially prepared for authors in one organization. It may be a field-specific manual like the American Chemical Society's *Manual for Authors and Editors*, or it may be a more general reference book like *The Chicago Manual of Style*. It may be a military specification or a publication standard developed by the American National Standards Institute.

Prepare a Style Guide

Whatever your writing task, begin with a clear idea of how finished pages will look and how the final document will be packaged. If you are working with coauthors, each member of the writing group needs the same instructions about the physical appearance of pages or screens. Individual authors will save time because they do not need to make style or format decisions. The resulting document will be consistent (Figure 7.1).

Think Visually

Visualize your completed document as a series of two-page spreads rather than a pile of individual sheets of paper. Two-page spreads are what readers will see when they read your document. Consider printing or photocopying on two sides. Such a method not only looks more professional but also increases the possibilities for placing figures and tables on the same spread of pages in which they are discussed. Number prefatory pages with small roman numerals, pages in the report body with arabic numerals, and pages in the appendix with alphabetic designators (Figure 7.2).

Design Individual Pages

Select a design for individual pages in the report body, and use it consistently, thereby helping readers to learn from the regularized placement of information elements (Figure 7.3). Will each page have a header? A footer? Will you print in one or two columns? Where will you place page numbers? Though traditional documents are $8\frac{1}{2} \times 11$ inches, consider smaller format manuals (typically 5×8 inches) for appropriate applications.

Specify Design Elements

To create visually effective documents, you'll need to make decisions about typography, white space, margins, and highlighting devices. Strong design decisions are based on a thoughtful appraisal of the audience for your document and their purpose for reading it. In some cases, research results from comparative studies in information design can provide guidance. Most designers agree, for example, that ragged right margins are more readable than justified margins, though justified margins have a more formal look. And most agree that readers have a hard time reading sustained copy printed entirely in capital letters.

Typography Select a type style and size for each textual element in your report and use it consistently. Textual elements include titles, headings,

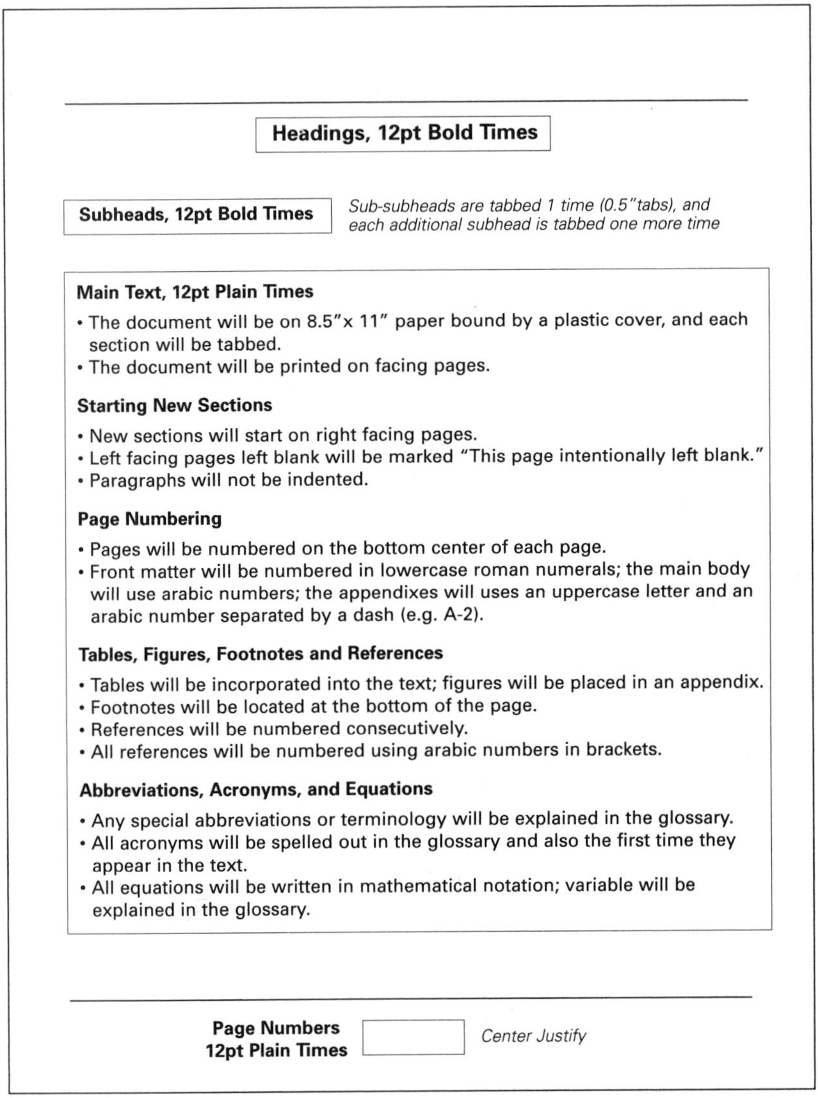

Headings, 12pt Bold Times

Subheads, 12pt Bold Times

Sub-subheads are tabbed 1 time (0.5"tabs), and each additional subhead is tabbed one more time

Main Text, 12pt Plain Times

- The document will be on 8.5"x 11" paper bound by a plastic cover, and each section will be tabbed.
- The document will be printed on facing pages.

Starting New Sections

- New sections will start on right facing pages.
- Left facing pages left blank will be marked "This page intentionally left blank."
- Paragraphs will not be indented.

Page Numbering

- Pages will be numbered on the bottom center of each page.
- Front matter will be numbered in lowercase roman numerals; the main body will use arabic numbers; the appendixes will uses an uppercase letter and an arabic number separated by a dash (e.g. A-2).

Tables, Figures, Footnotes and References

- Tables will be incorporated into the text; figures will be placed in an appendix.
- Footnotes will be located at the bottom of the page.
- References will be numbered consecutively.
- All references will be numbered using arabic numbers in brackets.

Abbreviations, Acronyms, and Equations

- Any special abbreviations or terminology will be explained in the glossary.
- All acronyms will be spelled out in the glossary and also the first time they appear in the text.
- All equations will be written in mathematical notation; variable will be explained in the glossary.

Page Numbers 12pt Plain Times *Center Justify*

Figure 7.1
This page format template creates standards for all documents produced at Cimarron Automation Services, Moorpark, CA. Note that the style guide itself is prepared in the recommended format.

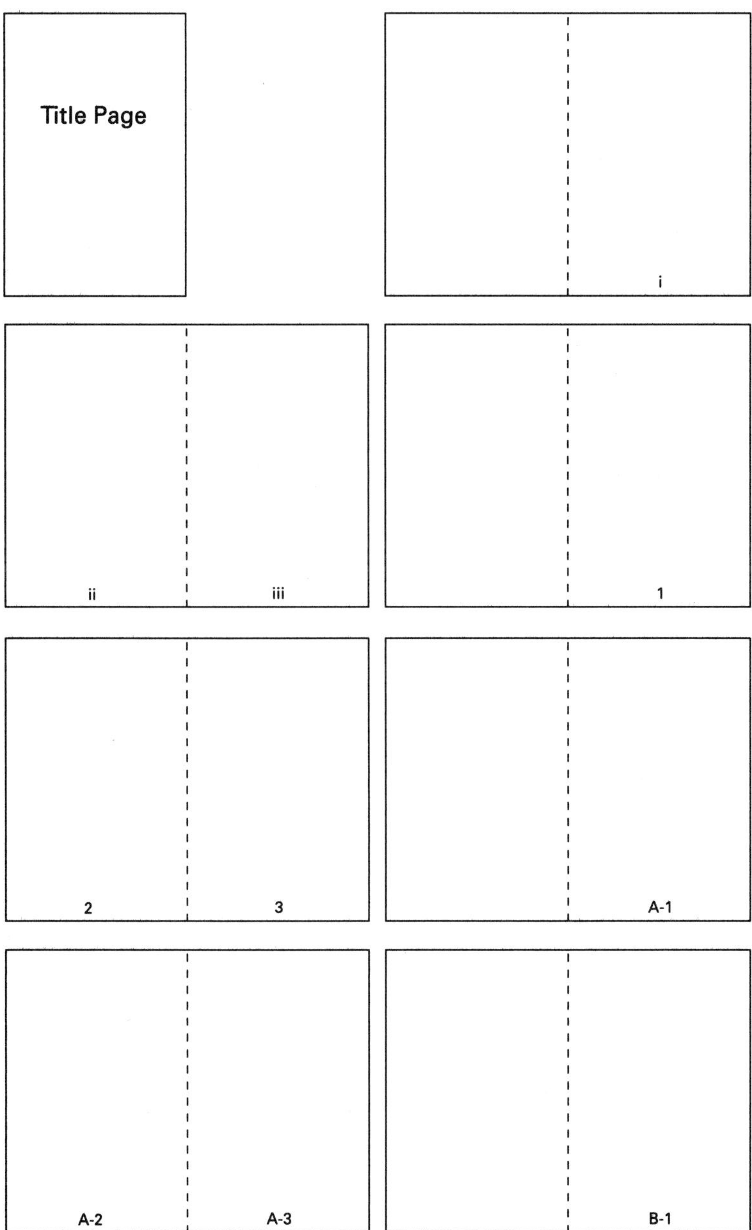

Figure 7.2
Plan documents as a series of two-page spreads, with expanded possibilities for placing illustrations and text on the same spread of pages. Note that prefatory, body, and appendix sections are clearly designated by the style of page numbers.

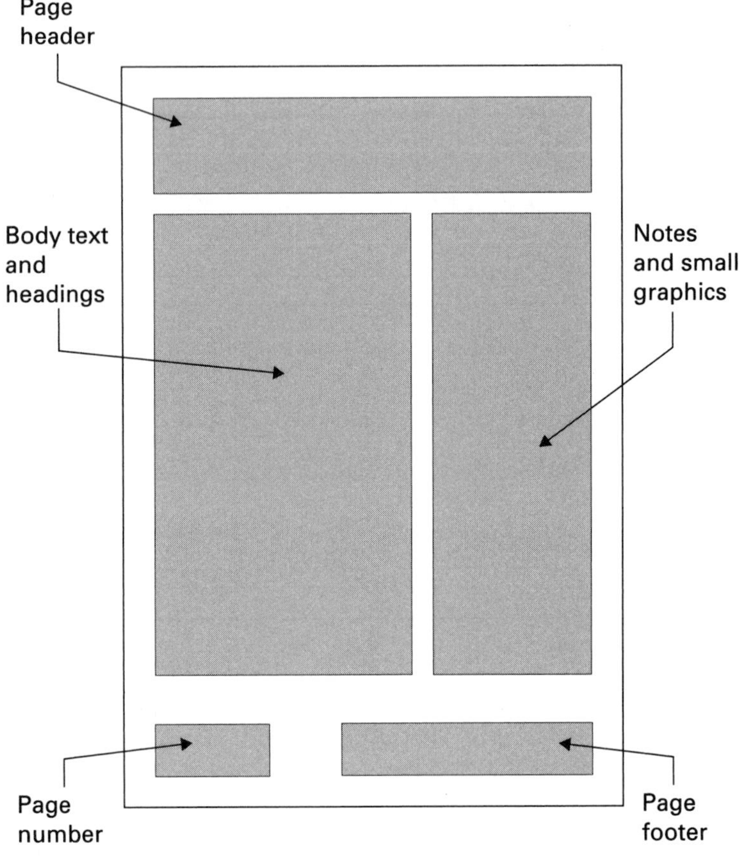

Figure 7.3
This page template would suit an instruction manual or a technical report with text, notes, and graphics. In this style, oversized illustrations are placed in appendixes.

text, captions for figures and tables, headers, footers, and references. Some typeface styles are easier to read than others. Many document designers recommend selecting a serif font rather than a sans serif for extended text. In serif font styles, small lines extend from the tops and bottoms of letters, apparently increasing readability. A 10- or 12-point type size for extended text is standard, though 12-point is easier to read. In general, write with upper- and lowercase letters. Extended text in uppercase letters is hard to read and best used for brief elements like headings (Figure 7.4).

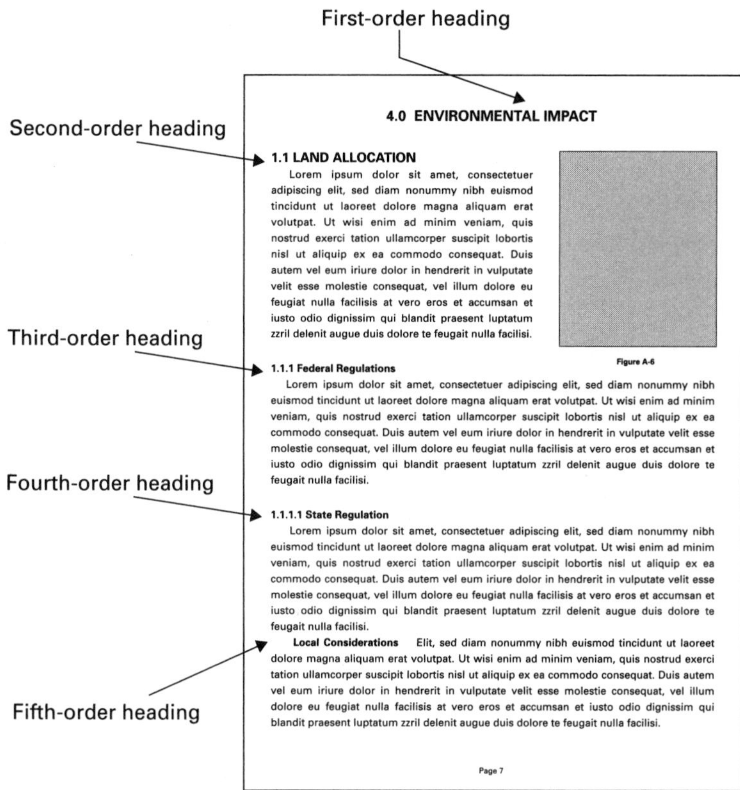

4.0 ENVIRONMENTAL IMPACT

1.1 LAND ALLOCATION
Lorem ipsum dolor sit amet, consectetuer adipiscing elit, sed diam nonummy nibh euismod tincidunt ut laoreet dolore magna aliquam erat volutpat. Ut wisi enim ad minim veniam, quis nostrud exerci tation ullamcorper suscipit lobortis nisl ut aliquip ex ea commodo consequat. Duis autem vel eum iriure dolor in hendrerit in vulputate velit esse molestie consequat, vel illum dolore eu feugiat nulla facilisis at vero eros et accumsan et iusto odio dignissim qui blandit praesent luptatum zzril delenit augue duis dolore te feugait nulla facilisi.

Figure A-6

1.1.1 Federal Regulations
Lorem ipsum dolor sit amet, consectetuer adipiscing elit, sed diam nonummy nibh euismod tincidunt ut laoreet dolore magna aliquam erat volutpat. Ut wisi enim ad minim veniam, quis nostrud exerci tation ullamcorper suscipit lobortis nisl ut aliquip ex ea commodo consequat. Duis autem vel eum iriure dolor in hendrerit in vulputate velit esse molestie consequat, vel illum dolore eu feugiat nulla facilisis at vero eros et accumsan et iusto odio dignissim qui blandit praesent luptatum zzril delenit augue duis dolore te feugait nulla facilisi.

1.1.1.1 State Regulation
Lorem ipsum dolor sit amet, consectetuer adipiscing elit, sed diam nonummy nibh euismod tincidunt ut laoreet dolore magna aliquam erat volutpat. Ut wisi enim ad minim veniam, quis nostrud exerci tation ullamcorper suscipit lobortis nisl ut aliquip ex ea commodo consequat. Duis autem vel eum iriure dolor in hendrerit in vulputate velit esse molestie consequat, vel illum dolore eu feugiat nulla facilisis at vero eros et accumsan et iusto odio dignissim qui blandit praesent luptatum zzril delenit augue duis dolore te feugait nulla facilisi.
Local Considerations Elit, sed diam nonummy nibh euismod tincidunt ut laoreet dolore magna aliquam erat volutpat. Ut wisi enim ad minim veniam, quis nostrud exerci tation ullamcorper suscipit lobortis nisl ut aliquip ex ea commodo consequat. Duis autem vel eum iriure dolor in hendrerit in vulputate velit esse molestie consequat, vel illum dolore eu feugiat nulla facilisis at vero eros et accumsan et iusto odio dignissim qui blandit praesent luptatum zzril delenit augue duis dolore te feugait nulla facilisi.

Page 7

Figure 7.4
In this heading style, a numbering system, upper- and lowercase typography, center placement, and bold type serve as markers of the hierarchical levels of information.

White Space The blank space in your document is an important design element. Used consistently, it improves readability and provides important information to readers about content. In creating a plan for using white space, include the size of margin around page perimeter as well as the amount of space you will leave between before and after headings, between paragraphs, and between major document sections. In the case of long documents, it is helpful to specify that each new section starts on its own right-facing page; the additional white space then serves as a physical break and reinforces the shift to a new topic. This strategy reinforces the modular organization of most technical documents while providing readers with natural places to stop and restart.

Highlighting Use document highlighting devices consistently but sparingly. Bold or italic type, color, and symbols such as arrows and boxes can provide helpful emphasis for readers, drawing attention to important points. Color has particular power to enhance information, though the cost of printing colored documents is often substantially higher than for black and white. Consider too that approximately 10 percent of adults are color-impaired, and do not rely entirely on color as a way to make crucial points.

Consider Options for Illustrations

You need a repertoire of strategies to deal with tables and figures. Full-page illustrations that can be studied without turning the page are commonly called "portrait" figures; illustrations that are wider than they are tall have a "landscape" orientation, requiring readers to turn the page sideways. When an illustration requires a landscape presentation, place it so that it can be viewed by rotating the page clockwise.

A figure or table too large to be contained in a portrait or landscape page can be made into a foldout, an oversized page folded to fit the dimensions of the printed report (Figure 7.5). Foldouts can be prepared with or without an apron, a blank page that forms the part of the foldout nearest the report binding. Aprons allow readers to open the foldout and refer to the illustration while reading the text, rather than alternating back and forth.

Include Navigation Aids

Help readers to navigate through your document. Tables of contents and indexes provide efficient directions to specific topics and are a welcome alternative to line-by-line reading. Headings provide previews of content and allow readers to skip directly to a desired subject. Page headers and page footers contain abbreviated information about the material on the page on which they appear: typically chapter and section titles and perhaps even an organizational logo. Divider pages, printed on heavier weight paper in a different color from the rest of the document, separate chapters and sections and provide additional navigation assistance. For long documents, specify that each divider page includes an expanded table of contents for the material in that section. Finally, tabbed section

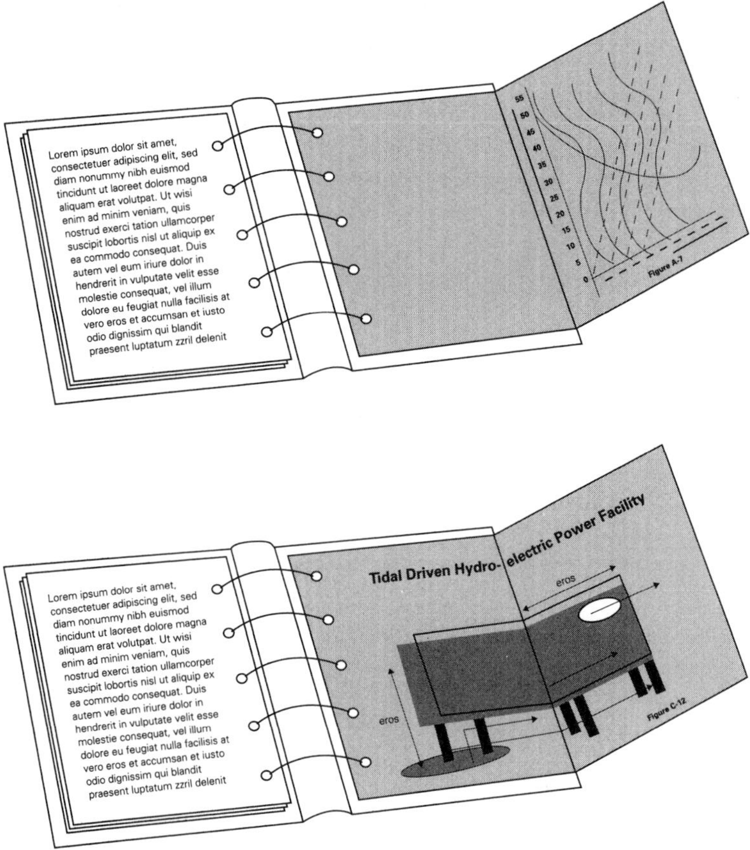

Figure 7.5
The top figure is presented as a foldout with an apron. Readers can refer to the figure while they are examining other pages of the report. The bottom figure is a standard foldout of an oversized illustration.

dividers help readers to navigate efficiently through your document (Figure 7.6).

Consider Binding Options

If you have choices about the final production of your document, remember that comb and spiral bindings help readers keep pages open and flat. Loose-leaf bindings are good for documents that must be updated with new pages. Heavyweight or plastic covers will help your document hold up through multiple readings.

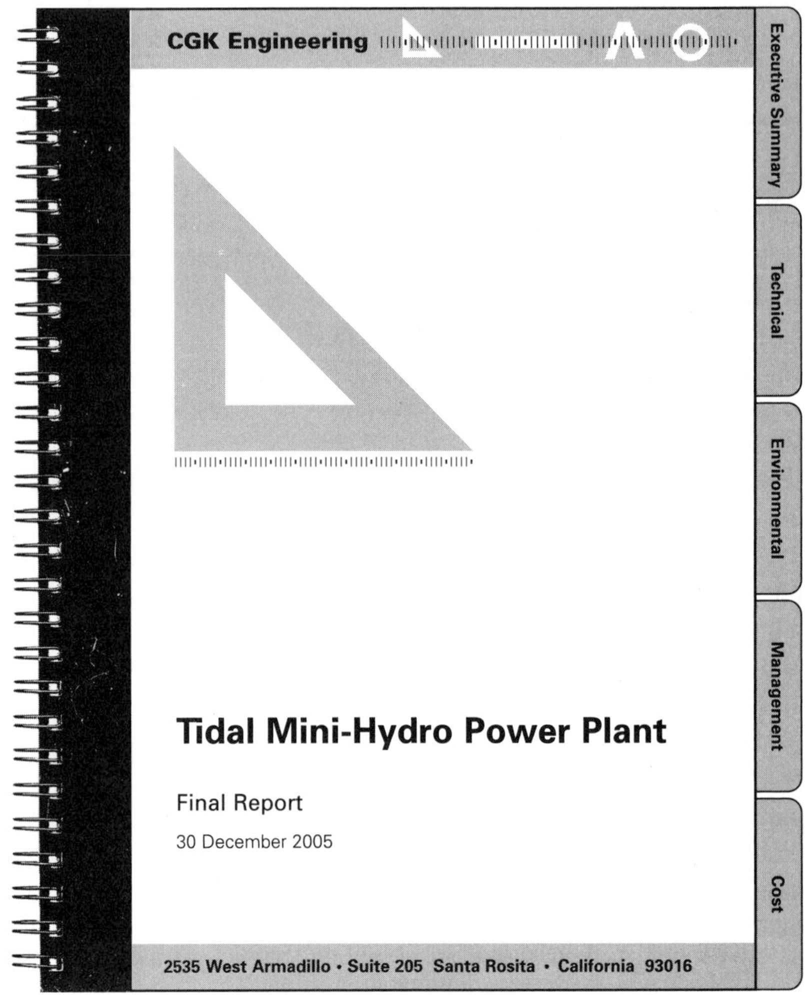

Figure 7.6
Labeled tabs serve as helpful navigation aids to readers.

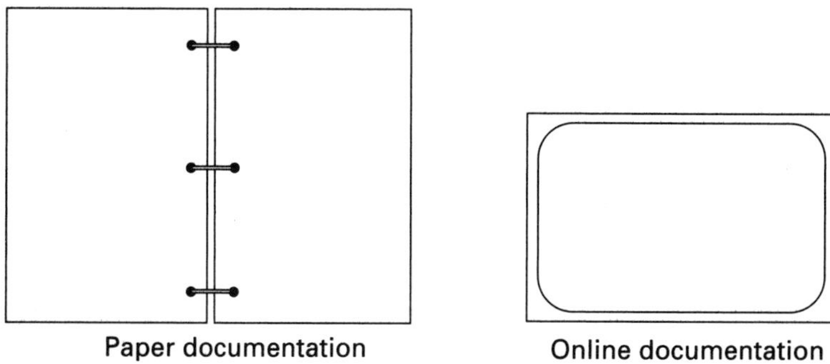

Figure 7.7
Readers are more familiar with paper-based formats. On-line displays are smaller and generally shaped in landscape rather than portrait orientation because of the shape of most computer monitors.

Electronic Document Standards

Many organizations achieve consistency of format, organization, style, and content by developing and using computer-based document templates. Writers are electronically constrained from violating standard formats. Though document templates do not eliminate all pains of authorship, they do answer such questions as what should be included, in what order, and at what level of detail. Document templates are frequently used in software development, where the need to document is urgent and sometimes overwhelming. A well-conceived set of templates provides guidelines for the eventual full set of documents, from the early specification stage to the shipping of the product.

Issues in Screen Design

Many design principles from the traditional world of ink-on-paper also apply to electronic documents. On a computer screen, consistency in design, generous use of white space, and avoidance of long text presented entirely in capital letters are as welcome as in hard-copy documents. But writers who publish information on computer screens also have urgent reasons to modify practices used in print.

Desktop and laptop on-line displays are smaller than pages (Figure 7.7), and information prepared for wireless delivery will be received in additionally reduced screen size. Text and graphics are grainier on screen than on paper, and reading on screen is more fatiguing than reading paper documents. To complicate matters, you cannot know how your on-line document will look on other platforms.

To overcome user's difficulties in reading electronic documents, develop a design for individual screens and use it consistently, exactly as you would for a print document. Define choices for typeface style and size for each text element. Consider using cascading style sheets to control and adjust the layout for an entire Web site with a single document that defines each style that will be used. Provide navigation aids that serve in the same way that elements like tables of contents, indexes, and running heads serve readers of print-based material. Well-designed computer screens can include titles, headers, maps, and even page numbers.

The Future of Document Standards

A static document, designed, printed, and bound in a certain way, is only one of many formats in which text can be presented. The term "document" now includes CD-ROM instructional modules with sound and animation as well as Web pages that may be read on the screens of handheld computers. New software makes it easier to convert text written for print delivery to HTML for Web delivery. Information can be prepared in such a way that it can be reused to serve a wide range of purposes, in varied formats and media. But achieving legible on-line text takes more effort than selecting a menu option that automatically converts a paper manual to a computer display. Despite advances in publication technology, authors still need to be aware of visual design features that facilitate reading.

8

Searching the Literature

■

You've just begun a new job in the R&D department of a chemical products firm. You were hired to be part of a new team that will eventually develop a new product line. Before your project begins, however, you need to identify current research that affects your team's plan. You're equipped with electronic access to a variety of databases, and you have access to an excellent research library, but you have only a few weeks to search the literature and summarize how it applies to your project. Where do you begin?

Literature searching always involves a time trade-off. Locating published information can support your research and may even streamline parts of your work. Staying abreast of developments avoids duplication of findings, but the published record is gigantic. Spending time tracking down information may not be useful. You can waste valuable hours better spent talking with colleagues and supervisors and working on your project.

The Flow of Technical Information

Searching the published record helps you

- Locate reference information necessary for routine research
- Gather data to extend your methods, findings, and discussions
- Follow the broad trends in your field and identify promising research problems
- Follow the theoretical, empirical, methodological, and design work of related fields

Remember that information in print is cold data. It's not likely to be up to date. The same scientific knowledge is released in many forms over several years. More current information flows orally among colleagues in a laboratory, passing through small in-house seminars and sometimes becoming bottled in proposals, progress reports, and memos, most of which are proprietary. In the private sector, the process often ends there, with limited circulation in print or electronic formats through a routing or mailing list. The information may also find its way into patents, specifications, manuals, and corporate bulletins.

In pure research, information moves outside the organization into various forums: conference proceedings, formal reports, and refereed articles. Moving from project initiation to journal publication takes up to five years. Four to eighteen months may pass as the manuscript goes through the review-editorial stage and emerges as a published article. Another month to a year may pass before the information is indexed and abstracted. Five more years may elapse before it is absorbed in reference works, including review articles and textbooks.

You can intercept information at several stages—in conversations, letters, seminars, colloquia, preliminary reports, theses, preprints, published reports, articles, literature guides, and reference works. Two factors vastly increase your ability to be current and concrete in your information searching:

- *Access to an expert.* The closer you get to the source of the expertise, the more current the information.
- *Electronic data.* On-line capabilities, both in local databases and on the World Wide Web, not only increase your reach, speed, and versatility in locating information but also give you access to information before it is in print. Increasingly, some materials appear only in electronic form.

The Reference Library

Libraries have three elements: catalogs, literature guides, and collections, or stacks (Figure 8.1). Catalogs, now mostly on-line, list the holdings of a library; they are your primary points of entry to what the library owns

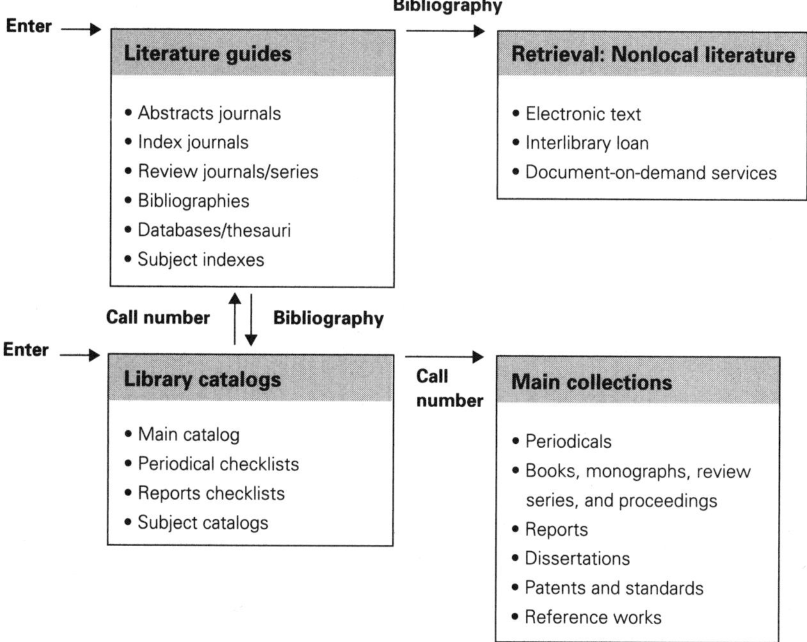

Figure 8.1
The library as a system. In contemporary libraries, the entire system, including the main collections, is increasingly linked together as a single on-line entity.

physically or has access to electronically. While catalogs list journals and other serial publications, they do not normally list individual journal articles, reports, or other short forms. Short publications are indexed in literature guides for specific fields. No library, however, will contain every item listed in guides like *Chemical Abstracts* or Compendex (formerly known as the *Engineering Index*).

Increasingly, catalogs, literature guides, and collections are incorporated in an electronically linked system. You enter the system from your workstation or personal computer, accessing either the on-line catalogs or the database of selected literature guides. From all these, you retrieve bibliographical entries and their call numbers (or URLs if for electronic access), which you then use to locate items of interest, either in hard copy in the library's collections or through electronic access. The texts held in the library's main collections may be hard copy, microform, electronic files, or compact disk technology (CD-ROM). CD-ROM storage increases space, with a single disk capable of storing up to 300,000 pages of print. The current trend in information retrieval is toward the building of large-scale digital libraries that are accessed on the Internet.

A typical on-line catalog entry for an author shows

- Author-title-subject information
- Publication and imprint information
- Call number
- Location and availability

For example, the on-line catalog screen shown in Figure 8.2 shows the search result for a volume titled *Lipids and Tumors*. On-line capabilities improve the speed and facility of library access. You can search through authors, titles, subjects, numbers (ISBNs, call numbers, government numbers, etc.), and keywords. Any of these categories can help you locate a desired text or help you concentrate sources of information.

Finding Technical Literature

You locate different kinds of documents by consulting general or specialized listings, including the main catalog and standard reference works. Consult reference librarians if you are unfamiliar with the guides that index the literature of your specialty.

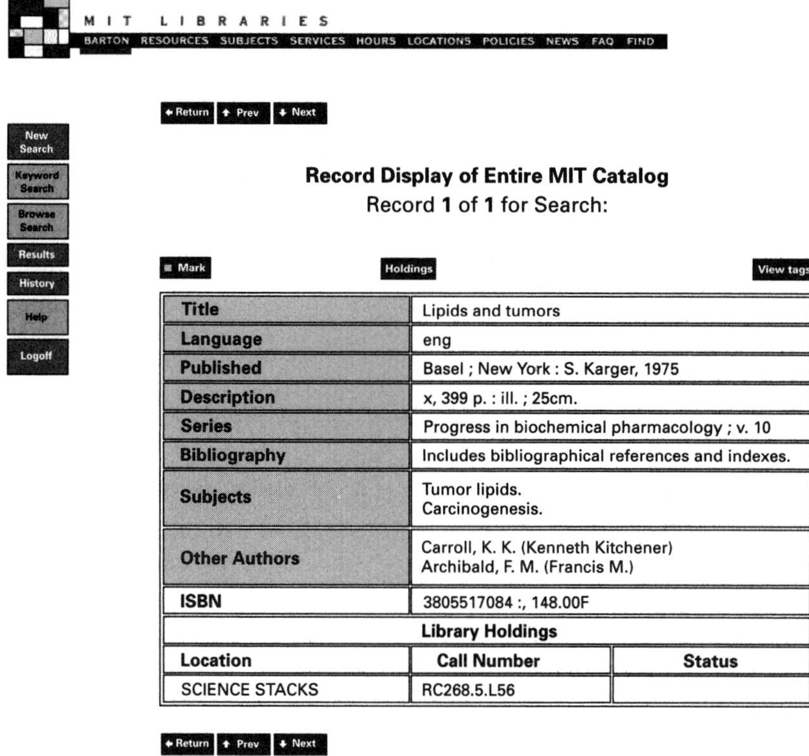

MIT LIBRARIES

BARTON RESOURCES SUBJECTS SERVICES HOURS LOCATIONS POLICIES NEWS FAQ FIND

New Search | Keyword Search | Browse Search | Results | History | Help | Logoff

Return | Prev | Next

Record Display of Entire MIT Catalog

Record **1** of **1** for Search:

Mark | Holdings | View tags

Title	Lipids and tumors	
Language	eng	
Published	Basel ; New York : S. Karger, 1975	
Description	x, 399 p. : ill. ; 25cm.	
Series	Progress in biochemical pharmacology ; v. 10	
Bibliography	Includes bibliographical references and indexes.	
Subjects	Tumor lipids. Carcinogenesis.	
Other Authors	Carroll, K. K. (Kenneth Kitchener) Archibald, F. M. (Francis M.)	
ISBN	3805517084 :, 148.00F	
	Library Holdings	
Location	**Call Number**	**Status**
SCIENCE STACKS	RC268.5.L56	

Return | Prev | Next

Return to Top

Figure 8.2
A typical on-line catalog entry for a title search. The same volume could be found under author, subject, keyword, call number or ISBN number.

Guides to the Literature

Literature guides list and abstract individual articles. Examples of these literature guides include *Physics Abstracts* and the *Engineering Index*, and their respective databases, INSPEC and COMPENDEX (see Table 8.1). These guides are indexed in the main catalog by title and corporate author (sponsoring organization). More than 2,000 abstracts journals cover the annual research output of the sciences and applied sciences. These literature guides are listed in various reference works, including C. D. Hurt's *Information Sources in Science and Technology* (1998), which arranges bibliographies and literature guides by field.

Abstracts journals and databases cover mostly articles and reports but also include patents, theses, proceedings, and books. Some, like the National Technical Information Service database (formerly, *Government Reports Announcements and Index*), are devoted to agency-sponsored research, usually issued as technical reports. Others, like *Computer and Control Abstracts* (INSPEC database) or the NASA Center for Aerospace Information Technical Report Server (formerly, NASA *Scientific and Technical Aerospace Reports* or STAR), are devoted to a field with many subdisciplines. The *Science Citation Index* and *Index Medicus* are interdisciplinary in scope and often overlap. If your library subscribes to the electronic version of a literature guide like *Chemical Abstracts* or a database like MEDLINE, you can search it electronically—possibly on the Web from your office computer.

The *Engineering Index Annual* (EIA), a sample entry of which is shown in Figure 8.3, is an abstracts journal. The EIA abstracts are arranged by subject, in keeping with *Engineering Information Thesaurus*. The entry in Figure 8.3 appears under the main subject heading "Biomechanics," and the subheading "Joints," after which abstracts appear in numerical sequence. If you have access to COMPENDEX, the on-line version of this index, you can search for authors, titles, keywords, and institutions (Table 8.1). Each abstracts journal or service is arranged differently, with its format described in an introductory section.

Journals

Professional journals are usually listed in the main catalog, which identifies their location and call numbers (e.g., Q.S399 for *Science Magazine*). Some libraries, especially those without on-line main catalogs, may list journals and other serial publications in a Periodicals Checklist located in a fiche file in the library's reference section. Many libraries maintain an on-line periodicals checklist, as well as an on-line listing linked to the electronic journals the library carries. When you find an article listing in a literature guide, you then go to the main catalog or periodicals checklist to find the journal location, usually alphabetically by title, and call number. For general information on content, addresses, publication schedules, and sponsoring agencies for journals in all fields,

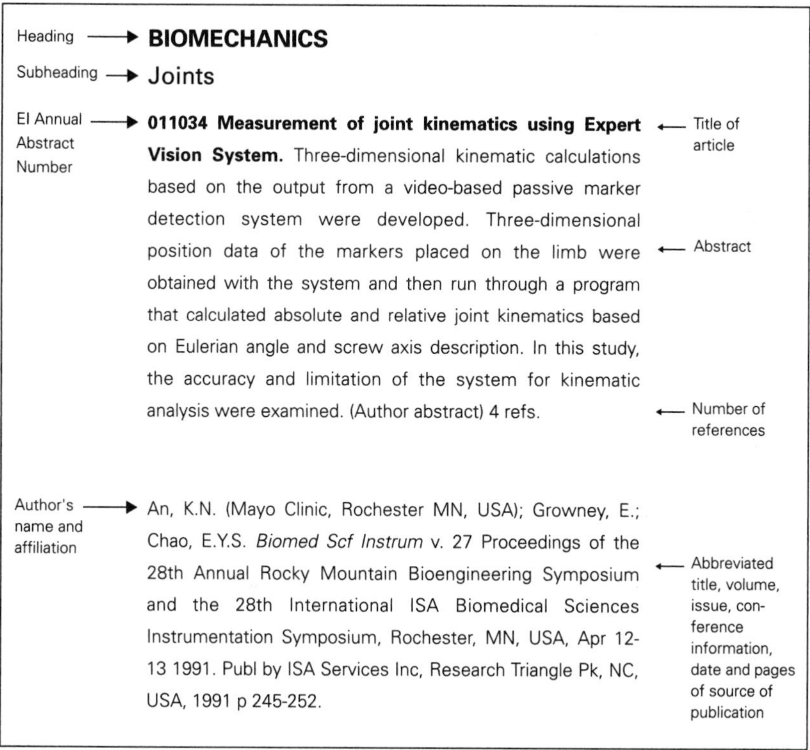

Amvrosova,0.1., 102487
An, Chae, 073109
An, K.N., 011034
An, Quang Dieu, 044888
Anada, Tetsuo, 109026
(a) Author index
Heading
BIOMECHANICS
Subheading
Joints
EI Annual Abstract Number
011034 Measurement of joint kinematics using Expert Vision System. Three-dimensional kinematic calculations based on the output from a video-based passive marker detection system were developed. Three-dimensional position data of the markers placed on the limb were obtained with the system and then run through a program that calculated absolute and relative joint kinematics based on Eulerian angle and screw axis description. In this study, the accuracy and limitation of the system for kinematic analysis were examined. (Author abstract) 4 refs.
Title of article
Abstract
Number of references
Author's name and affiliation
An, K.N. (Mayo Clinic, Rochester MN, USA); Growney, E.; Chao, E.Y.S. Biomed Scf Instrum v. 27 Proceedings of the 28th Annual Rocky Mountain Bioengineering Symposium and the 28th International ISA Biomedical Sciences Instrumentation Symposium, Rochester, MN, USA, Apr 12-13 1991. Publ by ISA Services Inc, Research Triangle Pk, NC, USA, 1991 p 245-252.
Abbreviated title, volume, issue, conference information, date and pages of source of publication
(b) Sample entry

Figure 8.3
Typical entry from an abstracts journal, *the Engineering Index Annual* (EIA): (a) author index entry, and (b) main listing. The entry is a conference paper, which is part of a volume in an annual symposium series (the Biomedical Sciences Instrumentation Symposium). EIA abstracts, which are part of the electronic database COMPENDEX, are arranged by subject, in keeping with the *Engineering Index Thesaurus*. (Used with permission.)

consult *Ulrich's International Periodicals Directory* in the reference section of your library. The Web resource ⟨jake.med.yale.edu⟩ is a useful tool for determining which databases index a particular journal.

Books, Monographs, Proceedings, and Review Series

Books of all kinds are listed in the main catalog under authors or editors, title, subjects, and corporate authors. Locating conference proceedings requires the conference title and date. If you need these, consult the librarian. Proceedings are also listed in the Institute for Scientific Information's (ISI) *Index to Scientific and Technical Proceedings* and in other literature guides such as COMPENDEX. Review series (e.g., *Advances in Bioengineering*) are listed under the series title.

Reports

Some reports are listed in the main catalog under author, corporate author, title, subject, or number. These entries, however, represent only a fraction of the report literature in a technical library. Normally, libraries maintain a separate reports checklist, either in hard copy or on-line, alphanumerically arranged by report number or government number. First, you locate a report and its number in a literature guide or database such as NTIS or the NASA Technical Report Server (⟨www.sti.nasa.gov.casitrs.html⟩). Then you identify its location by finding the report number in your library's reports checklist, identifying an outside vendor, or locating the report on a Web-based reports server. For example, the BNL series of Brookhaven National Laboratory is listed before the BNWL series of Batelle Pacific Northwest Laboratory, which in turn precedes the NASA series of the National Aeronautics and Space Administration.

Dissertations

In academic libraries, dissertations written at the same institution are indexed in the main catalog under author and title. Other dissertations in science and applied science appear in the *Dissertation Abstracts International, B, The Sciences and Engineering*. This reference work is usually available in the reference section of your library on CD-ROM, or on a Web link as the searchable database *Dissertation Abstracts Online*.

Standards and Patents

The vast literature of standards and patents is too diffusely distributed for all but a highly specialized library. Consult the reference librarian. Indexes to the patent and standards literature are often listed in the main catalog under subject headings like "Patents" and "Standards." ISI's *Derwent Innovations Index* is an electronically searchable citation and subject index for worldwide patent literature in the sciences and engineering. Two additional resources are the U.S. Patent and Trademark Office at ⟨www.uspto.gov⟩, which contains a list of regional depository libraries and a searchable patent database, and the Delphion Intellectual Property Network at ⟨www.delphion.com/ibm.html⟩ for access to the full-text and images of U.S., European, and Japanese Patents since 1974. To locate international standards literature from sector standards organizations, government agencies, and international standards organizations, contact the American National Standards Institute's NSSN database at ⟨www.nssn.org⟩.

Electronic Journals, Bulletins, and Discussion Lists

Electronic journals and discussion lists are proliferating. Some are useful; others are not. Electronic journals have become a major means of refereeing and disseminating research results. Many important journals like *Science* of the American Association for the Advancement of Science are now delivered in both hard copy and electronic formats and may be available on your local library network. Network bulletin boards and discussion lists provide access to ongoing technical discussions, conference announcements, job lists, news in the profession, software, language groups, and so on.

Conducting Your Search

Your search strategy depends on your task. Looking up a handbook to locate a physical constant differs from the months-long processes of reading, note taking, and building a comprehensive bibliography in a specific area. Always plan your search before you commit a lot of time to the process. If the search is not routine, you can save time and avoid becoming bogged down by consulting a research librarian and running

Table 8.1
Some key information vendors and their services (selections)

Vendor	Publications	Databases	Current Awareness	WWW:URL
Biosis (Biosciences Information Services) 2100 Arch St., Philadelphia, PA 19103	*Biological Abstracts* *BA/Reports, Reviews, Meetings* *Zoological Record*	Biosis Previews BIOSIS TOXLINE	×	www.biosis.org/
CAS (Chemical Abstracts Service) 2540 Olentangy River Rd, Columbus, OH 43210	*Chemical Abstracts* *CA Selects* *Chemical Titles* *International Coden Directory* *CAS Source Index*	STN International CAS Registry SciFinder Chemcats	×	www.cas.org/
DTIC (Defense Technical Information Center), 8725 John J. Kingman Rd, Suite 0944, Fort Belvoir, VA 22060	*DTIC Digest*	DROLS (Defense RDT&E online system) STINET Research Summaries	×	www.dtic.mil/
Engineering Information, Inc., Castle Point on the Hudson, Hoboken, NJ 07030	*Engineering Index*	Ei Engineering Village COMPENDEX	×	www.ei.org/

INSPEC (Information Services in Physics, Electrotechnology, Computers, and Control), IEE, Savoy Place, London WC2R OBL, UK	*Science Abstracts:* *Series A: Physics Abstracts* *Series B: Electrical and Electronics Abstracts* *Series C: Computer and Control Abstracts* *Electronics Letters*	INSPEC	×	www.iee.org.uk/
ISI (Institute for Scientific Information), 3501 Market St., Philadelphia, PA 19104	*Science Citation Index* *Index to Scientific and Technical Proceedings* *Current Contents* *Derwent Innovations Index* *CompuMath Citation Index*	Web of Science SCI SciSearch	×	www.isinet.com/
MEDLARS (Medical Literature Analysis and Retrieval System), National Library of Medicine, 8600 Rockville Pike, Bethesda, MD 20894	*Index Medicus*	MEDLINE (Medlars online) TOXLINE (Toxicology Information Online)	×	www.nlm.nih.gov/
NTIS (National Technical Information Service), US Dept. Of Commerce, 5285 Port Royal Rd, Springfield VA 22161	*Government Reports Announcements and Index*	NTIS NTIS E-ALERTS Federal Research in Progress AGRICOLA	×	www.fedworld.gov/

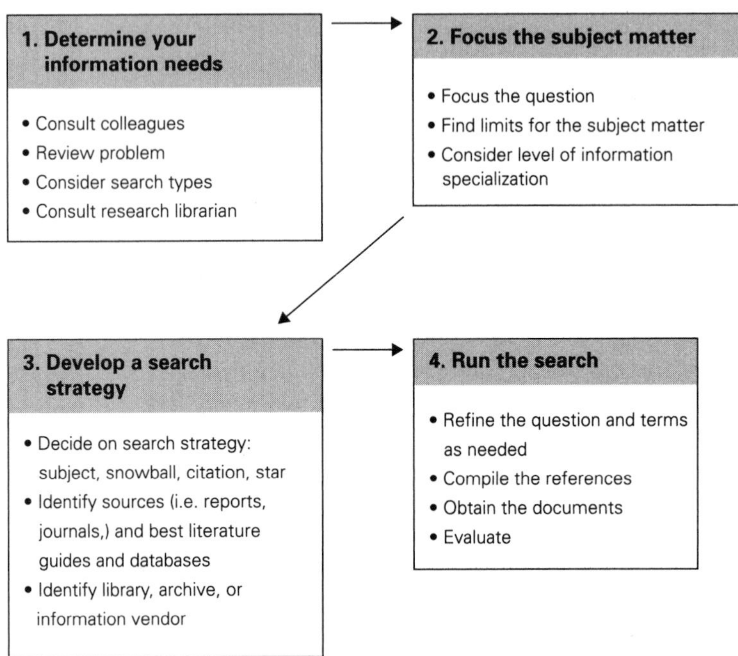

Figure 8.4
Steps in a search process.

a for-fee professional search. View literature searching in stages (Figure 8.4).

Determining Your Information Needs

Ask yourself what aspects of your problem might be explained in the published record. Think carefully about what you want to accomplish. You can spend hours flipping through catalog cards or wandering around electronic databases, hoping to find a useful listing. Much—maybe most—information will come to you from colleagues and by word of mouth. You can add to this information by focusing your search on a specific goal, like one of the following:

- *Bibliography search.* To fill out a citation or check its accuracy
- *Location search.* To retrieve a published item
- *Subject or concept search.* To isolate a class of information by using subject headings and keywords

- *Methodology search.* To find information about processes invented and refined by others
- *Follow-up search.* To trace developments in the theory, applications, or results of a field
- *Specific question search.* To find an answer to a specific question
- *State-of-the-art search.* To identify the most recent advances in theory or applications for a specific concept or process
- *Multidisciplinary search.* To concentrate information from sources across disparate fields
- *Comprehensive bibliography search.* To compile with the help of one or more databases an exhaustive list of sources treating a specific topic
- *World Wide Web search.* To search for information on the Internet using a Web search engine such as AltaVista, Google, or Yahoo!

A search can produce a single item on a computer screen or a massive listing of hundreds of items. If you are new to a field, you may need to read background literature so that you can talk intelligently about a topic. Start with broad sources and progress toward more specialized works (Figure 8.5).

Focusing the Subject Matter

Your search needs a focused question. Often, you can make a broad question more useful by rephrasing it. For example, the question "How good is the available underwater connector technology?" will produce more useful references if you rephrase it to ask "What was published on performance and reliability for underwater electrical connectors in 1992–2002?" The word "good" is now expressed as "performance" and "reliability," two key terms widely used in the field. "Underwater connector" is qualified by "electrical." The rephrased question also establishes time limits.

The less abstract and open-ended the question, the more likely you'll get concrete references. You can thus focus a search by refining your terms. If you are uncertain of the key terms for a topic, consult one of three widely used thesauri for science and engineering: *Thesaurus of Engineering and Scientific Terms* (Inspec 1995), *Library of Congress Subject Headings* (1975–), or the *Engineering Information Thesaurus* (1992–) of *the Engineering Index*. Most keywords and subject headings in on-line catalogs and databases are based on one of these three guides.

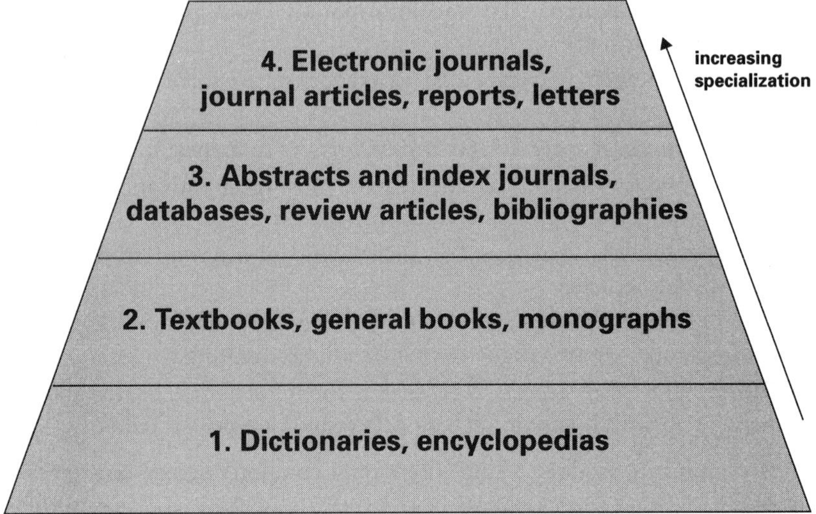

Figure 8.5
Increasing levels of specialization in studying a new field. If you are new to a subject area, you usually consult the broader references and then work upward.

Limit your search by considering the following:

- *Subject.* Key terms and subject headings
- *Sources.* Key publications, literature guides, or series
- *Time.* Inclusive dates for acceptable publications
- *Authors.* Specific authors or corporate sponsors of interest
- *Institutions.* Publications by individuals at key institutions
- *Documents.* Specific kinds of documents (e.g., patents, standards, reports)

You need not invoke all these limits in one search, but many literature guides and databases provide indexes and options that treat them all.

Developing Your Search Strategy

Searching for a specific publication is much easier than searching for general information. A specific publication requires you to locate a printed object. A general search requires you to concentrate information. When you have determined what you are looking for, you may decide just to glance at the indexes of a few relevant journals. You may also decide to pursue a more systematic strategy through literature guides and databases.

Ask colleagues and supervisors about literature guides and types of publications. Also check with reference librarians. For example, if research reports from the Naval Research Laboratory in Orlando, Florida, are possible sources, the NTIS database will be a useful literature guide. If articles in the journal *IEEE Transactions on Medical Imaging* are important to your work, you might want to see what abstracts journals abstract its articles.

As mentioned above, the Web resource JAKE (⟨jake.med.yale.edu⟩) is a useful tool for determining what databases index a particular journal. Information about where specific journals are indexed is also provided in *Ulrich's International Periodicals Directory* located in the reference section of your library. *Ulrich's* is available in hard copy, CD-ROM, and on-line formats (⟨www.ulrichsweb.com⟩). As Figure 8.6 shows, *IEEE Transactions on Medical Imaging* is indexed in more than a dozen different places. To trace a specific author, look in the author index of a literature guide. Electronic searching is versatile. Your query goes directly to the electronic files.

Subject Searching Subject searching means using subject headings or keywords to trace documents. With keywords, you can search either titles or subject areas. You identify keywords in books, articles, or thesauri. You then search the database, on-line catalog, or card catalog. Subject searches can be useful when you don't know much about the subject or when you just want to browse. Subject searching is also an excellent cross-disciplinary approach because the cross-references often show topical relationships between materials you don't normally associate.

In the sciences and applied sciences, however, the number of terms is so vast and expands at such a rapid rate that you need to use thesauri if you want to be accurate. In spite of its comfortable, encyclopedia-like feel, subject searching is often not the best technique. It's slow, even on-line, and it often produces barren lists of documents with little relevance to your interests. Subject searching on a Web browser can produce thousands of listings.

Sometimes, you can address this weakness by using review articles, reference guides organized by subject. Review articles identify and summarize key articles and other publications that have contributed to the

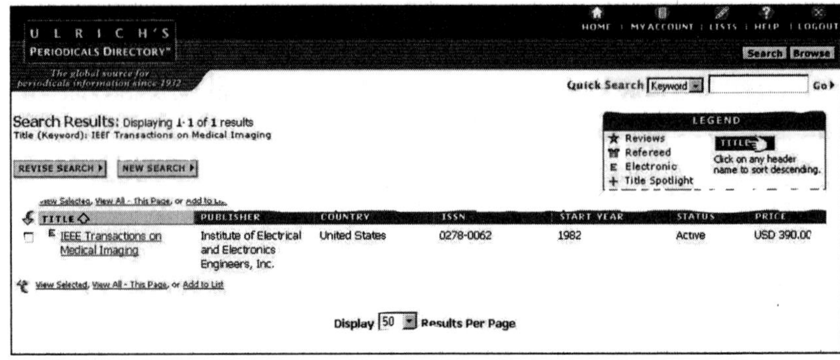

(a)

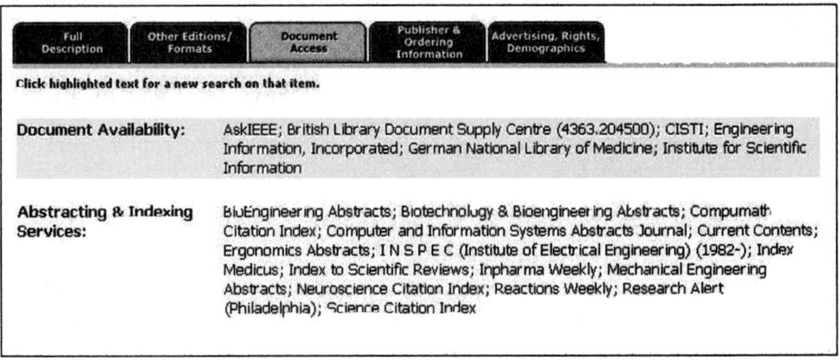

(b)

Figure 8.6
Web page entries for two ⟨ulrichsweb.com⟩ listings of the quarterly journal title, *IEEE Transactions on Medical Imaging*: (a) general title search results, and (b) document access. As shown in (b), more than a dozen abstracting and indexing guides cover the articles published in this journal.

development of a specific field. Some review articles are broad surveys of several years' work. Others are state-of-the-art accounts of annual or semiannual research. Review articles appear in standard refereed journals, in special review journals (e.g. *Chemical Reviews*), or in bound volumes with titles such as *Advances in* ... or *Progress in* ... Their unusually large numbers of references can help you identify review articles in literature guides.

Snowball Searching The most widely used searching technique, the "snowball approach," begins with a recent publication. You find a key paper, preprint, review article, or textbook. Then you look up items listed in the bibliography. From those retrieved items, you look up further entries, and so on, as Figure 8.7a shows.

The snowball search is fast and requires little use of literature guides. You assemble a list of references that supply you with more references. This technique does, however, have limitations. It tends to move you back to literature that is obsolete, and if you begin with a marginal article, you can spend much time assembling a network of similarly marginal papers.

Citation Searching In a citation search, you begin with a key source paper and compile a list of papers citing that paper. The basis of your search is that papers citing the source will be related. Just as the snowball search moves you backward, the citation search brings you forward because the papers citing are more recent than the paper cited, as Figure 8.7b shows.

The crucial literature guide for citation searching is the *Science Citation Index* (SCI) of the ISI (see Table 8.1). It is available in hard copy, on-line (*The Web of Science*), and on CD-ROM. The SCI consists of a citation index that lists the authors cited in footnotes and bibliographies of selected journals and books and a source index that lists authors of all citing publications. Covering a core of more than 5,700 journals, the SCI manages to account for a considerable percentage of the articles cited in basic science and applied science.

Citation searching offers distinct advantages. Like the snowball method, the citation search moves directly from document to document, with no intervening terminology or subject indexes. But the journals it

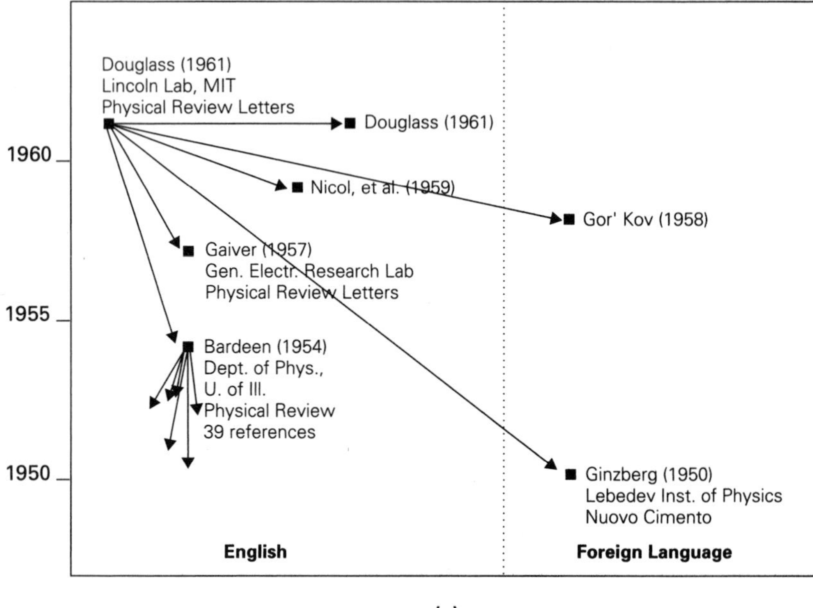

(a)

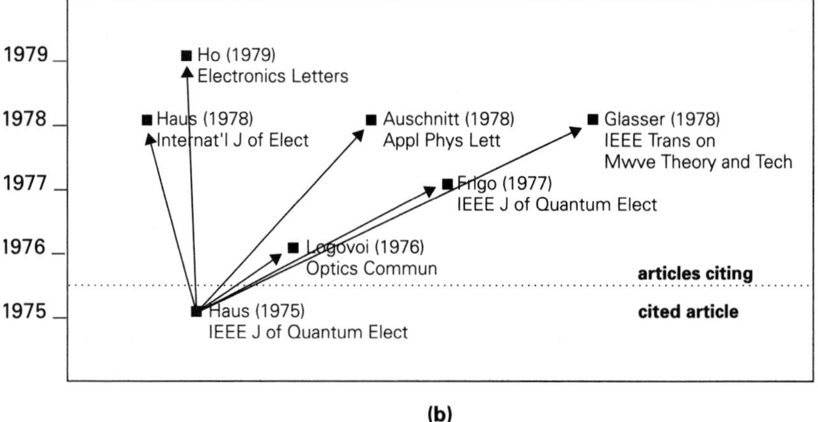

(b)

Figure 8.7
Two reference searches. The snowball search (a) moves the search back in time. An article by Douglass in *Physical Review Letters* (1961) takes you back to Nicol et al. (1959), Gaiver (1957), and Bardeen (1954). The article by Bardeen in the *Physical Review* is a review article with 39 references. In this way, you quickly find background literature and build a network of references. The citation search (b) moves you forward in time. Six different articles cite a source paper by Haus, which appeared in the *IEEE Journal of Quantum Electronics* in 1975. A more recent paper by Ho in *Electronics Letters* brings you forward to 1979.

covers are only part of the published literature. In addition, citation patterns are more significant in some fields than others. Citation searching has limited applications to engineering, private sector science, and foreign publication.

"Star" Searching With the "star" approach to searching, you look at certain star journals, researchers, or institutions that often account for a high proportion of the important publications and know-how in a specialized area. If you monitor their output, you will find critical work done in a given field. This approach offers focus.

To follow key journals, go to back issues, skim the article titles and abstracts, and glance over the cumulative subject indexes. To follow key authors, turn to the cumulative author indexes of journals or literature guides. You can also follow the research output of a facility known for its special work. For example, if a type of polymer research is being carried out in the materials science department of MIT or Stanford, you can track that research in the corporate index of *Chemical Abstracts*. Corporate indexes are standard features of many literature guides. Key authors and institutions may be searched together.

Comprehensive Database Searching You can carry out comprehensive database searches from a computer linked to one or more commercial database systems such as those listed in Table 8.1. You focus the search by using thesauri to develop a search profile of terms that will draw the desired information from the database. The *Gale Directory of Databases* is a semiannual publication which lists several hundred databases that cover many technical fields and special forms of literature such as reports, theses, and patents. Using these databases can be intimidating and time consuming. The operation may require the assistance of a reference librarian and may be conducted most effectively on a fee-for-service basis with the help of an information specialist. However, more and more user-friendly Web-based access is provided to support direct end-user searching.

Normally, the initial search profile needs to be tried out and focused until relevant titles and abstracts show on the screen. Add terms to narrow the search. Delete terms to broaden it. Database searching is useful when you need to compile a bibliography, gain access to restricted databases, or find recent bibliographical information before it appears in literature guides. The sheer time required to read and assess the results of this kind of search, however, can be significant.

Running the Search

Searches rarely go according to plan, so you need to combine planning with initiative. You might begin, for example, with one or two papers furnished by a colleague, retrieve selected references listed in those papers, and then refer to a literature guide that indexes journals in which the most interesting papers appear. The key terms in the literature guide may help you modify your own list of keywords. If you find a valuable author, you might look in the literature guide's author index to see what else that author has published. Examine the corporate index to see if the author's research group is producing other papers on the subject. You might also glance at the SCI to see what other publications have cited your key papers.

When you have compiled a list of potential papers, you need to retrieve them. This process can be time consuming. Examine the abstracts of the papers carefully to see if you really need the complete work. Papers on topics close to your own research may be especially important if you're in danger of duplicating the research of others. Photocopy and collect important papers. Note other references of limited interest. Keep your notes up to date.

Current Awareness Searching

Because advances and innovation are the lifeblood of scientific professions, staying informed is critical. The best way is through routine, informal oral exchanges with colleagues, as well as peer group discussions at professional and trade conferences. Formal current awareness searching, by contrast, can keep you abreast of developments outside your circle of acquaintances. It also requires special literature guides.

One simple but useful tool is the contents journal, which prints the contents pages of recent professional journals. *Current Contents*, a series of small biweekly contents journals published in hardcopy and electronic formats by the ISI comes in different specialized fields. One is the *Physical, Chemical, and Earth Sciences* Edition. Another, the biweekly publication of Chemical Abstracts Service, called *CASelects*, covers over 1,000 journals. All the information vendors listed in Table 8.1 offer extensive current awareness services, which are described at their respective Web site URLs.

A related service is the published search. Many information vendors publish prepared searches, each based on a set of keywords that identifies specific subjects of interest. These bibliographies may be ordered directly from the vendor (see Table 8.1).

Information Vendors

Information is now a commodity, and big information vendors offer a variety of services. The most specialized (and expensive) are the Selective Dissemination of Information (SDI) services, which tailor searches to the individual purchaser. SDI services, offered by most information vendors, periodically match a database to search keywords provided by the purchaser. The results are mailed or transmitted electronically. These services are also available for new patents, government standards, and military specifications.

Available services include

- Published reference works such as abstracts and index journals, bibliographies, and other compilations
- Current-awareness services, including contents journals and SDI services
- General and specialized databases available for on-line searching
- Document-on-demand acquisition services by e-mail, telephone, and mail

Services available from nine information vendors are listed in Table 8.1. Each vendor supplies literature and maintains a Web site that outlines specific services. Two even larger information vendors that market thousands of information products, including many of those listed in Table 8.1 and elsewhere in this chapter, are Ovid Technologies (⟨www.ovid.com⟩) and Dialog, a subsidiary of the Thompson Corporation (⟨www.dialog.com⟩).

9

Documenting Sources

■

You have agreed to write the literature review for the proposal your work group is about to finish. Your job will be to summarize the earlier research that has led the principal investigator to frame the question and apply for research support. You know the science involved, and you're familiar with 10 years of technical literature. The content should be easy. But now that you begin to write, you wonder how much to quote, how much of the summary is so basic that it needs no references, and what reference style to use.

Proposals, reports, and journal articles almost always contain references to earlier published work. If you have done a good job of researching your subject, the quality and quantity of the sources you have consulted will enhance your work. Your work acquires credibility when you review the literature and show that your contribution extends from a solid foundation of respected research.

In citing sources and preparing an accurate reference section, you make it possible for readers to retrace your steps by locating and consulting the papers that informed your work. They may wish to assess your work in the light of previous contributions to the subject, or they may be reading for ideas to help them with new or continuing projects.

Preparing Citation and Reference Sections

In reporting the results of a literature review, you have two tasks. For one, you need to write *citations* to the earlier work you are reporting. These citations appear in the text, at the place where you have referred to the earlier work. For the other, you must prepare a *reference section* or *bibliography* with specific details about authorship, publication, editions, and dates for each citation. Reference sections appear at the end of the document.

In scientific and engineering fields, variants of two citation styles and at least one hundred reference styles are commonly used. The appropriateness of each is determined solely by the expectations of potential readers. Use whatever style you choose consistently throughout a document, but know that the particulars of style are established by convention and group practice. Even within specific fields, however, a variety of styles may be permitted. The American Chemical Society, for example, allows three forms of citations in ACS publications (superscript numbers; italic numbers on the line, in parentheses; author name and year in parentheses).

Find out if style guidelines are available in your work setting or provided by your anticipated audience. If you are submitting an article for journal publication, check the journal for advice about preferred reference style. Consider purchasing the style manual of your professional society. The American Chemical Society, American Mathematical Society, American Medical Association, American National Standards Institute, American Institute of Physics, and the Council of Biology Editors are some of the groups that publish guidelines for authors. Many of these organizations provide extensive information on their Web sites.

Several software packages are available for managing references. You create a database entry for each item, storing bibliographic information such as author, title, date, publisher, place of publication, volume, number, and page. When you prepare the citations and reference section, you can choose from a menu of styles, and the software will automatically format your paper. If you revise your paper for a publication with different documentation standards, the software will reformat as necessary.

Citing References in Text

In scientific and engineering writing, two broad styles are commonly used for citing earlier work. In one style, the author's name and year of publication are printed in the text, enclosed in parentheses (see Figure 9.1a). The other style is based on a number, sometimes set on the line in parentheses or square brackets, sometimes set in superscript format (see Figure 9.1b). If the number appears in superscript format, it usually (but not always) refers to a footnote or endnote rather than to an item in a reference list. In each case, the citation refers to more complete bibliographical information provided in a footnote or at the end of the document in the reference section.

If you are free to select a style for your in-text citations, consider these comparative features. Many readers prefer author's name/year of publication citations in a literature review. When names and dates are embedded in the text, readers immediately know what researchers you have consulted and can assess the currentness of your search. Numbered references do not intrude on the text, so they are easier to skip over if you are reading strictly for research findings. If they are handled sequentially, though, numbered references may present problems when you add or delete anything.

Preparing a Reference List

The precise bibliographic form for items in your reference list will be established by the style guide you select. For author's name/year of publication citations, references appear in alphabetical order. For numbered citations, references usually appear in the order in which you have referred to them, though some journals prefer alphabetical order in these cases as well.

Author's Name and Year of Publication

Citation in Text

After extrapair paternity, divorce has recently attracted perhaps the greatest amount of attention in both theoretical and empirical studies of partnership in birds (Black 1996). Divorce is ultimately related to mate choice and provides one of the best examples of the value of using game theoretical models in the study of sexual selection (Ens et al. 1996; McNamara et al. 1999).

References

Black MJ (1996) Partnerships in birds: the study of monogamy. Oxford University press, Oxford

Ens BJ, Choudhury S, Black MJ (1996) Mate fidelity and divorce in monogamous birds. In Black MJ (ed) Partnerships in birds: the study of monogamy. Oxford University Press, Oxford, pp 344–401

McNamara JM, Forslund P, Lang A (1999) An ESS model for divorce strategies in birds. Phil Trans R Soc Long B 354: 223–236

Source: S.M Ramsay, et al. 2000 Divorce and extrapair mating in female black-capped chickadees (*Parus atricapillus*): Separate strategies with a common target. *Behavioral Ecology and Sociobiology 49: 18–23.*

Figure 9.1a
Author's name and year of publication. In this citation style, the author name(s) and year of publication appear in parentheses in the body of the paper, and each citation links to a reference section at the end of the paper. Within this general style, you will find a range of variations. Some journals require square brackets rather than parentheses; some require a comma between name and date; some list first author only; some list up to three authors.

Numbered Citation

Citation in Text

A recent method for producing fullerenes is the synthesis of soot, which contains fullerenes, by the laser-pyrolysis of gas-phase hydrocarbons [1, 2]. This method is based on the interaction between the radiation from a CO_2 laser and the gaseous reactants in a cross-flow cell [3].

References

1. *Ebrecht, M., Faerber, M., Rohmund, F., Smirnov, V., Stelmach, O., and Huisken, F., Chem. Phys. Lett. 213:34 (1993).*

2. *Voicu, I., Armand, X., Cauchetier, M., Herlin, N., and Bourcier, S. Chem. Phys. Lett. 256:261 (1996).*

3. *Haggerty, J.S. Cannon, W.R. in Laser-Induced Chemical Processes (J.I. Steinfeld Ed.), Plenum Press, 1981, New York, Chap. 3.*

Sources: S. Petcu and M. Cauchetier. 2000. Formation of fullerenes in the laser-pyrolysis of benzene. *Combustion and Flame* 122: 500–507.

Figure 9.1b
Numbered citations. In this citation style, each number refers to an item in the final reference list. Numbers are almost always enclosed in square brackets rather than parentheses, to distinguish citations from equations.

If you have no other instructions to follow, some typical bibliographic forms are provided below. Note that personal communications and other forms of "nonrecoverable" information such as interviews or phone conversations are usually not included in the reference section, though they may have been mentioned in the body of the article. Distinguish multiple items published by one author in the same year with lowercase alphabetic letters: (Wilde, 1997a; Wilde, 1997b).

Book. McClintock, F. A. and A. S. Argon. 1966. *The mechanical behavior of materials*. Reading, MA: Addison-Wesley.

Journal article. Ulmer, B. and H. Ishii. 2000. Emerging frameworks for tangible user interfaces. *IBM systems journal*. 39: 3:915–931.

Article in an edited collection. Engelbart, D. C. 1960. A conceptual framework for the augmentation of man's intellect. In *Computer-supported cooperative work*, edited by I. Greif. San Mateo, CA: Morgan Kaufmann Publishers.

Report. Winward, A. H. 2000. Monitoring the vegetarian resources in riparian areas. Report No. RMRS-GTR-47. Ogden, UT: U.S. Department of Agriculture.

Dissertation or thesis. Cavallo, D. 1996. Leveraging learning through technological fluency. Master's thesis. Massachusetts Institute of Technology Media Laboratory.

Conference paper in proceedings. Fettwels, A., and M. Nossek. 1981. Sampling rate increase and decrease in wave digital filters. *Proceedings, 6th IEEE symposium on circuits and systems*, 839–841. Chicago, IL: Institute of Electrical and Electronic Engineers.

Patent. Gershenfeld, N. September 21, 1993. Method and apparatus for electromagnetic non-contact position measurement with respect to one or more axes. U.S. Patent No. 5,247,261.

Standard. American Society for Testing and Materials (ASTM). 1979. Standard for metric practice. PCN 06-503807-4l.

Citing Sources for Tables and Figures

If you photocopy, scan, or otherwise reproduce a figure or table from another publication for inclusion in your document, you must credit the source. Even if you redraw the illustration but borrow significantly from the original, you need to cite the original author and publication. References to figures and tables can be identified by the word SOURCE (usually

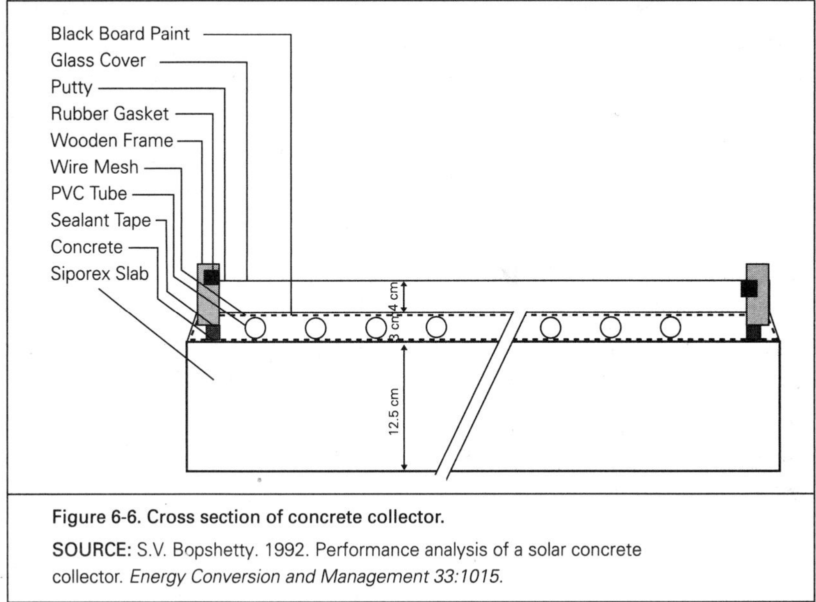

Figure 9.2
When tables and figures are reproduced from their original sources, the source is identified directly under the caption. In most cases, the original caption needs to be rewritten for its new context.

in small capital letters) and treated as an integral part of the artwork (Figure 9.2).

Citing Electronic Information Sources

Though technology has advanced more rapidly than bibliographic practice, reference conventions have emerged for information gleaned from electronic media. The purpose of any reference section remains the same: to enable readers to locate and examine all documents referred to by the authors, including those that exist only as computer files.

In citing an electronic publication, your goal is to help readers to retrace your steps if they choose. Title, author, and date the material was written are crucial pieces of information, as they are in hard copy. It is helpful to include the publication medium (on-line database or CD-ROM, for example) and the name of the vendor or on-line publication service. But you will have to be flexible in the information you provide.

Details are not always available. Though a scrupulous Web site should include a page with month and day when it was last updated, such information is not always provided. For that reason, some researchers also include the date when they accessed the site referred to which they refer. In general, personal e-mail is considered "nonrecoverable" information and is not included in a reference section, though it may have been mentioned in the text.

Abstract of a journal article from a database. Longo, N. and S. Langley, L. Griffan, and L. Elsas. May, 1995. Two mutations in the insulin receptor gene of a patient with leprechaunism. *Journal of clinical endocrinology and metabolism*, 80 (5) [Online abstract]. Medline Plus, National Library of Medicine. Item: UI 95263703.

Article from an electronic journal. Bhandarkar, S. V., and S. H. Neau. December 15, 2000. Lipase-catalyzed enantioselective esterification of flurbiprofen with n-butanol [On-line journal]. *EJB Electronic Journal of Biotechnology*, 3 (3). Available: http://www.ejb.org/content/vol3/issue3/full/3/.

CD-ROM. Aquatic Sciences and Fisheries Abstracts 1988–1994. [CD-ROM] Cambridge Scientific Abstracts. Compact Disc SP-160-010. Silver Platter. Available: http://www.silverplatter.com.

Computer program. *EndNote* 4 (Word for Windows 2000 compatible format). [Computer program]. ISI ResearchSoft. Available: http://www.endnote.com.

Web site. V. Gerasimov and W. Bender, Swings that think. Retrieved April 21, 2001 from the World Wide Web: http://vadim.www.media.mit.edu/stt/bat.html.

Paraphrasing and Quoting Ideas from Sources

Even in documents explicitly designated as literature reviews, technical authors do not usually quote extensively from their sources. Rather, they summarize key points, restating ideas in their own words. However, whether you quote exactly and enclose the borrowed phrases or sentences in quotation marks or you restate and summarize in your own language, you must still credit the source. Only when the ideas and information are considered common knowledge can you omit a citation. A simple test of whether ideas are common knowledge is this one: Would this idea or piece of information be familiar to someone with your academic and professional standing (perhaps a colleague) who has not researched the subject? If the answer is yes, you do not have to cite the source. Otherwise, you must indicate the source of the material, even if it appears in several texts.

Paraphrases

When you paraphrase, you restate ideas in new forms that are original in both sentence structure and word choice. Taking the basic structure from a source and substituting a few words is not an acceptable paraphrase and may be construed as plagiarism. Similarly, creating a new sentence by merging the wording of two or more sources is also unacceptable.

For the long direct quotation presented in Figure 9.3, the following restatement is not acceptable as paraphrase because it is too close to the original: Electric-field nulling reduces unwanted hot spots in or on the target body; electric-field focusing maximizes the delivered RF power. These goals can be achieved by using an adaptive hyperthermia phased-array system (Fenn and King 1992, 236).

Here is an acceptable paraphrase:

During heating of a tumor site, an adaptive hyperthermia phased-array system can achieve two important clinical goals: reduced incidence of unwanted hot spots and improved power delivery (Fenn and King 1992, 236).

Note that both the sentence structure and some of the wording have been changed, so that quotation marks are unnecessary.

Short Direct Quotations

If a quotation occupies four lines or fewer in your manuscript, incorporate it in your text and use quotation marks to indicate where your words stop and the quotation begins. In the author/date format, provide a page number within the parentheses:

Both electric-field nulling, which reduces the "occurrence of undesired hot spots inside or on the surface of a target body," and electric-field focusing are "intended to maximize the RF power delivered to a tumor site" (Fenn and King 1992, 236).

xxxxxxxxxxxxxxxxxxxxxxxxxxxxxxxxxxxxxxxxxxxxxxxx
xxxxxxxxxxxxxxxxxxxxxxxxxxxxxxxxxxxxxxxxxxxxxxxx
xxxxxxxxxxxxxxxxxxxxxxxxxxxxxxxxxxxxxxxxxxxxxxxx

> The purpose of electric-field nulling is to reduce the occurence of undesired hot spots inside or on the surface of a target body. In contrast, electric-field focusing is intended to maximize the RF power delivered to a tumor site. The data presented in this article suggest that both these goals can be achieved with an adaptive hyperthermia phased-array system (Fenn and King 1992, 236).

xxxxxxxxxxxxxxxxxxxxxxxxxxxxxxxxxxxxxxxxxxxxxxxx
xxxxxxxxxxxxxxxxxxxxxxxxxxxxxxxxxxxxxxxxxxxxxxxx
xxxxxxxxxxxxxxxxxxxxxxxxxxxxxxxxxxxxxxxxxxxxxxxx

Source: A.J. Fenn and G.A. King. 1992. Adaptive nulling in the hyperthermia treatment of cancer. *Lincoln Lab Journal* 5: 223–240.

Figure 9.3
Long direct quotations are set off from the text, without additional quotation marks.

Long Direct Quotations

If a quotation is five lines or longer, set it off from the text by beginning a new line and indenting one inch from the left margin as shown in Figure 9.3. Do not use quotation marks. If you are using author/date citation style, provide an exact page reference.

Partial Direct Quotations

Use ellipsis marks, usually three spaced periods, to indicate any deletions you have made in the quotation:

> Fenn and King have demonstrated "electric-field nulling . . . to reduce the occurrence of undesired hot spots inside or on the surface of a target body . . . and electric-field focusing . . . to maximize the RF power delivered to a tumor site" (1992, 236).

Altered Quotations

Use square brackets to mark any alterations you have made to the quotation. In the following example, the author is quoting directly from Fenn and King's paper, but she has added to the original source an explanation of the abbreviation RF:

> Fenn and King have demonstrated "electric-field nulling . . . to reduce the occurrence of undesired hot spots inside or on the surface of a target body . . . and electric-field focusing . . . to maximize the RF [radio frequency] power delivered to a tumor site" (1992, 236).

Managing Citations and References

Researchers frequently reuse reference sections in continuing projects, building on earlier library and database searches as they enter new phases of their work. If a paper is rejected by one journal, they may submit a revised version to another journal with different documentation requirements.

Maintain *complete* bibliographic information about each cited item so that you can modify a reference section if necessary, without spending valuable time looking up details like beginning and end page numbers or the full names of coauthors. Develop a process for keeping it all together, whether you use software specially designed for such tasks, adapt a database program for the purpose, or create your own scheme for maintaining complete bibliographic information.

Reference sections often interest readers long after the research methods reported in the article have been superseded by new approaches. A good reference section provides access to the history of a technical problem. Your references can be as valuable as your research methods and findings.

Part II

10

Memos, Letters, and Electronic Mail

■

As the newly hired project manager for a software design team, you spend your first days on the job reading the files. You have little time to interview team members and ask follow-up questions. By the end of your first week, you must report to your manager with a detailed proposal and work plan for a stalled project. Your assessment of both the history of the project and the roles played by the engineers in your group will be based on your review of archived e-mail. Team members' correspondence will form your first impressions.

Memos and letters are brief and relatively informal documents, yet many technical professionals spend more time writing (and reading) these familiar forms, in hard copy or as electronic mail, than they spend on any other communication task. Despite their brevity and relative informality, memos and letters may be archived and reviewed later, often by those not originally addressed. Both forms may become important parts of a project record. They may serve as the basis for important decisions, with effects as significant as those of multivolume proposals or articles published in prestigious journals.

The structure of both memos and letters is flexible enough to be useful for a wide variety of purposes, including proposals, requests for information, trip reports, complaints, inquiries, records of telephone conversations, or calls for meetings. The personalized forms of memos and letters distinguish them from other technical workplace documents. They name the recipient, they name corecipients, and they identify the author.

In recent years, e-mail has dramatically increased the correspondence workload for many engineers and scientists. E-mail has blurred distinctions between the traditional memo and letter forms, opened new communication channels, and changed the way that information flows in many organizations.

The Difference between Memos and Letters

The memo form is used for communicating *within* an organization, never for an outside audience. The letter is used for communicating *outside* an organization. Thus a feasibility report prepared for exclusive use within a company will be accompanied by a *memo* of transmittal, and a report prepared for a client will be covered by a *letter* of transmittal. Social practices will vary, of course. A supervisor wanting to congratulate an engineer for having a paper accepted for publication might send a letter to a home address, rather than a memo through the interoffice mail.

In e-mail communication, no distinction is made between memo and letter or between files that will be transmitted to the next office and files that will be transmitted across the country or around the world. E-mail written to a colleague in the next office looks exactly like e-mail written to a client on another continent. Gone are the social signals and organizational images communicated through letterhead. You, under your user name, write to someone else with a user name. All user names are more or less the same length, without clues to educational or professional status. Most e-mail recipients open their own mail—even those who never read hard copy memos or letters until a secretary has opened envelopes and logged in each document. Most e-mail readers answer their own mail—even those who otherwise dictate copy for secretarial transcription.

Reaching Your Audience

In shaping the content of memos and letters, you must address the information needs of your recipient. In your search for a persuasive strategy, consider what your reader already knows about the situation you are addressing. Ask yourself how this reader is likely to react to what you are saying. Then remember that the first audience for memos and letters may not be the last. If copies of your document need to be sent to other readers, you should also consider how each one is likely to respond to what you have written.

Brevity and Focus

Though memos and letters are frequently many pages or screens long, we recommend using these correspondence forms for brief accounts of single issues, with a goal of one-subject, one-page (or one-screen) for each document. The subject should be specified in the subject line, and the content should relate to the stated subject. For two subjects, write two documents. In that way, each subject can receive your reader's full attention, and each document can be appropriately filed for retrieval at a later date. Realistically, the conventional format of letters requires so much space for formalities that it is often difficult to hold to a one-page limit. Nonetheless, we recommend brevity.

Design for Emphasis

For hard copy memos and letters, visual presentation is crucially important: memos look like memos; letters look like letters. But faithfulness to outward appearance is not enough to ensure effective communication. Simply following a prescribed format will not help you to write a memo or letter that suits its particular context.

Though your memo or letter may be brief, do not assume that every word will be read with interest and rapt attention. Ask yourself how you can best design your page for a reader who may not read straight through or who may spend only a minute or so skimming what you have written. Make judicious use of bullets, numbered lists, headings, and bold type to emphasize the ideas you want to get across. Remember that you are competing for the attention of readers who probably have too much to read and too much to do. The burden of calling attention to key points rests with you, not with your reader.

Memos

Memo Format

Though the exact placement of elements in the heading of memos will vary from organization to organization, the content remains constant: memo headings invariably identify date, recipient, author, and subject (Figure 10.1). Memo headings perform important reference functions. The prominence of the date provides a chronology for the issue under consideration, so anyone can see at a glance where each document fits into the evolving life of a project. The date locates each action and may be important later if, for example, you are involved in legal action. Organizational titles and levels of responsibility may influence the relative weight a reader will give each communication. Although scientists and engineers should be influenced primarily by objective evidence, readers are, nevertheless, often influenced by the professional rankings of authors and audiences.

Of all the elements of a memo, the subject line carries most responsibility for flagging readers. Because it functions as title and abstract combined, the subject line needs both to present a concise statement of the memo's topic and to contain information that will tell a reader whether the memo is immediately important. An additional audience for the subject line is the clerical personnel who file your document. They are likely to make filing decisions based on mechanical searches for keywords. An

Internal Correspondence General Specifics Research Corporation

DATE: 5 April 2004

TO: Marion McSputz
Director, Research and Development

FROM: Ryan Ashker
Staff Scientist

RE: FUNDING REQUEST: CHARACTERIZING AND MAPPING WETLANDS IN AMAZONIAN SAVANNAS

I am requesting a budget of $135,000 to develop a strategy for characterizing and mapping seasonally inundated wetlands in Amazonian savannas. The savanna wetlands of the Amazon Basin are not well understood, and they appear to have a significant impact on global biogeochemical cycles as well as a key role in maintaining global biodiversity.

In order to develop a strategy for characterizing and mapping seasonally inundated wetlands, I propose to do the following tasks, over a period of two months:

1. Determine which currently operating satellite sensors can be used to detect inundation. My evaluation will be based on the spatial, temporal, and spectral resolutions of the sensors.
2. Determine the amount of in situ fieldwork necessary to characterize seasonally inundated savanna wetlands.
3. Use river stage and precipitation data to determine the timing of low and high water periods.

I have attached a detailed budget as well as maps of the wetlands area. Please call me at extension 3-3090 if you would like to discuss this project.

Enclosures (2)

Figure 10.1
This memo heading contains four requisite elements: date, name of recipient, name of sender, and subject. The subject line is focused and specific; the body of the memo, with its numbered list, is designed for rapid reading. (Courtesy of Ryan Ashker.)

Generic Subject Lines	Informative Subject Lines
Proposal	Proposal to Develop Melanin-Like Polymer as Passive Solar Energy Converter
Problem	Shipping Delay Stalling Development of Optical Components
Colloquium	Colloquium on DNA and Quantum Computing, 11 August 2005, MIT

Figure 10.2
Informative subject lines contain concise statements of the memo subject, giving readers a helpful preview of content. By comparison, generic subject lines do very little to address the information needs of potential readers.

ambiguous subject line can keep your memo from reaching the right reader at the time you send it and later on as well (Figure 10.2).

Memo Organization

Though the external forms of memos and letters are rigid, the content is extremely malleable. Once you identify your purpose and audience, you can shape your text more precisely than for other technical documents. Each memo or letter you write should adhere to some broad outlines, but within those outlines you develop strategies for organizing and presenting your content to a specified audience.

A three-part organizational plan works well for most memos. Open with an overview. Tell readers exactly why you are writing and what they will gain from reading. Use the middle section of the memo to develop your point and provide supporting arguments. Use the final section to summarize your point and, when appropriate, to request or suggest follow-up action. Consider adding internal headings to give your reader a quick preview of contents (Figure 10.3). If your memo is more than one page, include a heading that will allow your document to be reassembled if pages become separated (Figure 10.4). Always indicate the presence of attachments or enclosures with a notice at the bottom of the page.

Memo Style

Memos are utilitarian forms, less formal than letters. In most organizations, memo writers initial their documents in the heading and do not

Internal Correspondence General Specifics Research Corporation

DATE: 5 April 2004

TO: Marion McSputz
Director, Research and Development

FROM: Ryan Ashker RA
Staff Scientist

RE: FUNDING REQUEST: CHARACTERIZING AND MAPPING WETLANDS IN AMAZONIAN SAVANNAS

PROPOSAL

I am requesting a budget of $135,000 to develop a strategy for characterizing and mapping seasonally inundated wetlands in Amazonian savannas.

WORK PLAN

In order to develop a strategy for characterizing and mapping seasonally inundated wetlands, I propose to do the following tasks, over a period of two months:

1. Determine which currently operating satellite sensors can be used to detect inundation. My evaluation will be based on the spatial, temporal, and spectral resolutions of the sensors.
2. Determine the amount of in situ fieldwork necessary to characterize seasonally inundated savanna wetlands.
3. Use river stage and precipitation data to determine the timing of low and high water periods.

CONCLUSION

I have attached a detailed budget as well as maps of the wetlands area. Please call me at extension 3-3090 if you would like to discuss this project.

Enclosures (2)

Figure 10.3
The internal headings in this version of the memo presented in Figure 10.1 provide helpful information about the structure of the argument. (Courtesy of Ryan Ashker.)

Second page of a memo

Characterizing and Mapping Wetlands
5 April 2004

Figure 10.4
If your memo is longer than one page, include an identifying heading on subsequent pages.

sign their full names. But memos are also personal: by all means, use "I" and "you." A memo is an internal document, and formality is not expected. Aim for a style that is efficient and cordial.

But keep in mind that despite their in-house status, memos may become important parts of historical archives. You may be tempted to include a private communication in technical memos; for example, you may want to use the occasion of reporting progress on a new stack gas emission control to add congratulations on the birth of a baby. Yet the personal rarely seems appropriate months or years later. Remember that your memo may need to be reviewed. Many writers attach removable notes to memos and use those spaces for personal comments that they would not want retrieved at a later date.

Letters

Letter Format

Most organizations have a "house style" for letters, with standards for indentation, spacing, and punctuation. The widely used block style is both attractive and functional (Figure 10.5). Though a subject line is not absolutely required, it provides a preview for the recipient and filing information for an assistant who may need to retrieve the letter at a later date. Some organizations prefer modified block. In this style, paragraphs are indented, and date, closing, and signature are aligned approximately two-thirds across the page. As with memos, be sure to put identifying information on second pages and to indicate the presence of enclosures.

Letters should always be addressed to someone, never to "Dear Sir" or "To Whom It May Concern." If you do not know the name, title, and preferred form of address of the person you're writing to, you should

STEPHEN CHUNG AND ASSOCIATES

PROFESSIONAL CONSULTING ENGINEERS
18886 Hollister Ave.
Santa Rosita, California 93069

6 December 2006 *4 returns*

Mr. Nicholas Altenbernd, President
Quantum Design
18 Wright St.
Cambridge, 012139 *2 returns*

Dear Mr. Altenbernd: *2 returns*

SUBJECT: Topic Agenda for Year-End Progress Meeting

____
_____________________________________.

____
____
____________________:

- ________________________________
- ________________________________

____
____________________________. *2 returns*

Yours sincerely, *4 returns*

Kenneth Manning
Director of Research *2 returns*

km/db *2 returns*

Enclosures (2) *2 returns*

c: David Neuman

Figure 10.5
This left-justified style with subject line is a functional form for letters. The informative subject line provides a helpful preview of content, and a bulleted list will be used to highlight main points. The recipient knows to expect two enclosures in addition to the letter, and he is informed that a copy will be sent to another person. The initials of the assistant who helped to prepare the letter ("db") are also included.

not, except in unusual circumstances, be writing a letter. Check details with care, and do not assume goodwill. Most people are irritated when their names are misspelled or their titles garbled.

Letter Organization

No all-purpose form letter will achieve the results you want for all occasions, for all readers. Like memos, letters must be designed to reach the specific reader named as recipient, the specific readers named as co-recipients, and unknown readers who are likely to read the document at some later date.

The recommended three-part organization for memos works well for most letters. Open with an overview, telling the reader exactly why you are writing. Use the middle section of the letter to develop your point. Use the final section to summarize your point and to suggest follow-up action. Use typographical and page design features to highlight key points.

Though the middle sections of technical letters are closely related to the spare and utilitarian style of memos, the openings and closings are strictly ceremonial. Letter writers are more constrained than memo writers to make verbal gestures that are purely social.

Letter Status

A letter is simultaneously highly personal and official. You speak directly to the intended reader with the salutation "Dear," and you close the document with your handwritten signature. At the same time, the letter may bear your company letterhead and highlight your administrative level. A word processor's initials at the bottom of the page will signal to your reader that you are important enough to have secretarial assistance. And when you include the title and organizational address of your recipient, you indicate that your letter is both written and received in full recognition of institutional hierarchies.

Letters written on organizational letterhead are official forms, and they relay the weight of your office and affiliation. Because communication on company letterhead carries an implied official endorsement, take care when you use it. You are, in effect, expressing not only your own message but also the views of your organization.

Electronic Mail

Reaching Your E-Mail Audience

While e-mail is a supple instrument for sharing ideas and information, the volume of e-mail in networked writing environments frequently leads to cognitive overload. As a result, e-mail messages are often just skimmed, not scrutinized carefully. A closely related problem with e-mail is that few readers are willing to read extended on-line text. Important e-mail is often printed out or followed up by a conventional memo or letter.

If you want your e-mail messages to be read, you will have to consider that the recipient of your message may be receiving dozens of messages along with yours. With most e-mail systems, the person to whom you are writing will receive a list of mail to read, identifying the author and displaying the subject line. Nothing obliges a recipient to retrieve and read what you have sent; in most e-mail systems a user can delete unwanted mail without reading it. Ignoring e-mail is as easy as scanning the return address on an unopened envelope and dropping the entire piece of hard-copy mail in the nearest trash basket.

As a writer, you naturally want to increase the likelihood that the person to whom you have written will read your message. Try to alleviate cognitive overload by writing a straightforward, information-dense subject line. Keep your message brief: one screenful for one message. Use page design features like bulleted and numbered lists, as you would in hard copy (Figure 10.6). Achieve and maintain credibility: Don't send junk e-mail, tempting as it is to take advantage of the ease with which distribution lists can be expanded and text, graphics, Web pages, audio, and video files can be attached to your message.

Evolving Conventions

E-mail can function as either memo or letter. When you correspond on paper, you follow well-known conventions about whether to write in memo or letter format. With e-mail, you need to make some decisions on your own, often mixing practices depending on your relationship with the recipient of your e-mail and your purpose for writing. When you write to people outside of your own organization, it is helpful to include an e-mail "signature" at the bottom of your message, with your full

VERSION #1

Date: Fri, 10 Nov 2004 13:16:51 -0800 (PST)
From: Anne Foster <foster@marinescience.gsrc.org>
To: Henry Yang <yang@marinescience.gsrc.org>
Subject: Re: Cooperative Oceanographic Monitoring System

We have often talked about our need for an efficient system of collecting, analyzing, and disseminating the information gathered by our oceanography research groups. I propose that we investigate a new communications network that enables collection and collation of research data. This network is known as a cooperative oceanographic monitoring system (COMS). By implementing a COMS, we would be able to offer our researchers and clients more reliable research, and accurate modeling. We would also increase communication between research departments, facilitate rapid collation and retrieval of information, and have the potential to link effectively with international research stations. Please contact me if you would like to pursue this idea.

....

Anne Foster, Department of Marine Science
General Specifics Research Corporation FAX: (805) 893-8651

VERSION #2

Date: Fri, 10 Nov 2004 13:16:51 -0800 (PST)
From: Anne Foster <foster@marinescience.gsrc.org>
To: Henry Yang <yang@marinescience.gsrc.org>
Subject: Re: Cooperative Oceanographic Monitoring System

We have often talked about our need for an efficient system of collecting, analyzing, and disseminating the information gathered by our oceanography research groups. I propose that we investigate a new communications network, a cooperative oceanographic monitoring system (COMS). By implementing a COMS, we would achieve the following advantages:

- Synchronized data collection, detailed analysis, and accurate modeling
- Improved communication between research departments
- Faster collation and retrieval of information
- Increased potential to link effectively with international research stations.

Please contact me if you would like to pursue this idea.

....

Anne Foster, Department of Marine Science
General Specifics Research Corporation FAX: (805) 893-8651

Figure 10.6
Compare these two versions of an e-mail message. You can improve the readability of electronic mail by using lists and headings to emphasize key points. Note that the author provides additional contact information in a "signature" appended to her message.

name and additional relevant contact information, as shown in Figure 10.6. When you write to people with whom you do not have ongoing relationships, it is courteous to open with a salutation ("Dear Professor Banerjee"), as you would in a hard copy letter.

Some e-mail authors are comfortable with more forceful expression (called flaming) and less meticulous grammar and spelling than they would ordinarily use in hard-copy memos or letters. Such stylistic informality may not be appreciated. In corporate settings, where mail goes to many people on large mailing lists and is often forwarded and cross-posted, chances are that someone with a low tolerance for grammatical and spelling errors will receive your message. Always assume that verbal restraint and careful editing are valued qualities in professional settings.

The Status of E-Mail

E-mail is a technology in cultural transition, appearing to flout much time-honored company, university, and laboratory practice connected with hard copy memos and letters. When e-mail addresses are made public, correspondents tend to overstep conventional boundaries created by organizational hierarchies: 65 employees may write to one supervisor, altering long-held conventions about who writes to whom. In networked university settings, many professors note that students are more willing to ask for help with assignments through e-mail than in face-to-face meetings or by telephone.

Much of what happens for both writers and readers of e-mail is constrained or made possible by software design. Most e-mail systems present writers with a template: date and author's name are already filled in; names of others who should receive copies of the message are easy to insert. Even the subject line may be preformed (for example, "Reply to your message of 9/16"). Most templates have no space for anyone's title. You don't need to know whether your recipient has been promoted from Associate Director to Director or whether she prefers being addressed as Professor, Dr., Ms., or Mrs.

But nothing in electronic communication prevents it from becoming a form with rigid and elaborate social signals. Just as readers of hard copy can quickly size up the importance of a message by noting the organizational name and address on the letterhead and the writer's name and title, e-mail templates may be redesigned to provide recipients with social cues to indicate which files can be safely deleted before reading and which files need immediate and careful attention. As the volume of e-mail becomes overwhelming, e-mail recipients create lists of system users from whom they do not want to receive communication, and they request unlisted electronic addresses.

The legal status of electronic messages is complex and ambiguous. Some organizations are openly monitoring e-mail, and employees have been dismissed for what an employer considered inappropriate or unprofessional comments. Increasingly, e-mail messages, including those assumed to have been erased, are used as evidence in criminal and civil lawsuits. Other cases involving privacy and access are unresolved. E-mail users will do well to write cautiously in this environment, not mixing the personal and the professional.

Memos and Letters as Part of a Continuum

Your memo or letter may not be the last words on a subject. Your document may create additional communication tasks, and its relevance may extend well beyond any time frame you can imagine. Create electronic files of memos and letters for future reworking into additional documents. Most e-mail systems provide filing and storing options, though some e-mail users prefer to download important documents.

Finally, do not be overly dependent on writing as a method for communicating ideas. Be prepared to talk on any subject you have written about. The response to your memo or letter may include telephone calls and face-to-face meetings, both formal and informal. In the work of science and engineering, a written document is rarely the only form through which you will communicate with others.

11

Proposals

Proposals as Sales and Planning Documents
Proposals as Persuasion
Proposals as Projections
Strategic Planning for Funding Success
Solicited or Unsolicited?
Enter the Right Competitions
Think in Two Time Frames
Take Advantage of Help
Use Evaluation Criteria as Planning Tools
Learn about the Review Process
Get Approvals in Advance
Systematic Proposal Preparation
Study the Request for Proposal
Turn Requirements into Outline and Compliance Matrix
Make Plans for Time Management
Allocate Team Responsibilities
Prepare Style and Format Guides
Facilitate Electronic Submission
Proposal Content
Front Matter
Body of Proposal
Appendixes
Business Plans
Stressing the Strengths of Your Ideas
Preparing an Attractive Document Package
Your Proposal Writing Program

■

As the manager of a research team, you know that the future of your department depends on a steady stream of funds. You spend part of every month scanning agency releases for possible new requests for proposals. So far, you've been awarded enough contracts to sustain your research, but budget constraints have now forced you to limit the size of your staff. No longer can you afford to have a documentation manager available to coordinate, compile, and complete the many proposals that your department submits. You and your staff will have to assume more of the responsibility for proposal writing. And you'll need to find a way to accomplish this task while research continues.

For many engineers and scientists, proposals are the most important form of writing. Most academic research—and a substantial amount of industrial research—is funded through a review procedure in which written proposals are evaluated by panels of researchers from the same field. For working scientists and engineers, proposal writing can make the difference between continued research and interruption in a long-term project.

Proposals set projects in motion and are often part of a cycle of documents that marks the progress of research. They may be preceded by a preproposal called the *white paper*, an information package describing new concepts or products. In many technical industries, white papers are posted on a company Web site or mailed to prospective clients, in hope of receiving a request to provide the items described. The work specified in a proposal may be tracked in notebooks and progress reports. Memoranda, reference papers, meeting minutes, and letters then keep a project in motion.

Proposals as Sales and Planning Documents

Proposals are written in a variety of informal and formal modes, from short memoranda to multivolume industrial bids. An in-house proposal, written as a brief and informal memorandum, may circulate only within a writer's organization. An external proposal may circulate widely and be refereed by management and budget experts as well as by knowledgeable technical specialists. Business plans, written to acquire funding for a technical project, are a specialized form of proposal, typically submitted to an audience that includes bank loan officers and investors as well as company management.

Depending on the complexity and extent of a research project, a proposal may be written by one or by many researchers. For large industrial proposals, the production group may include, in addition to engineers and scientists, technical managers, editors, text processors, artists, and photographers.

Despite these differences, most proposals have important elements in common. They identify a problem; explain what work will be done to solve the problem; name the researchers who will do the work; argue for their qualifications; specify a time frame, location, materials, and equipment; and calculate a cost. Most proposals are submitted to reviewers who are knowledgeable, critical, and concerned, interested in selecting strong proposals and eliminating problematic ones. Many proposals have multiple reviewers; the more you are asking for—the higher the stakes—the more readers you are likely to have, and the more knowledgeable and critical they will be.

Proposals as Persuasion

A major difference between proposals and other forms of scientific and technical literature is that proposal documents are usually entered into competitions. The goal of every proposal writer is to win the approval and the money to go ahead with a project. Because success in preparing proposals is a major factor in advancing or even maintaining academic careers, as well as staying in business, writers must overcome any reluctance to draft persuasive documents.

Proposals are mixed bags of elements—technical descriptions, time lines, curricula vitae, budget analyses, fill-in-the-blank data sheets, and more. Think of ways to make every element in a proposal an argument for the value of your idea, the elegance and good sense of your work plan, the strength of your preparation, the appropriateness of your facilities, and the economy of your budget. But a successful proposal requires more than technical details. It requires a narrative shaped to exhibit the strengths of your plan. A well-developed proposal shows that the investigator has grasped a problem well enough to justify second-party sponsorship of the enterprise.

Proposals need just the right sales pitch: the goal is to get a sponsor to spend money. The usual strategy of academic proposal writers is to understate claims, trying to sound somewhat reticent and modest, cautious and competent. In contrast, the usual strategy of industrial proposal writers and authors of business plans is to aggrandize. The industrial attitude, with a frank and forthright interest in winning contracts, may seem overly aggressive to academics. In truth, however, all proposal writers seek to win. Keep that goal in mind, whatever the differences in appropriate decorum.

Proposals as Projections

A proposal is much more than a pitch for financial support. It is a planning document that defines work commitments and establishes the criteria by which the success of a project will be determined. You write a proposal before you know the results. But the illusion that a proposal must foster in its reviewers is that it represents work for which there is already a plan. Estimates of a work program, its cost, and its schedule must be convincing. The effective proposal writer must be imaginative, able to convert ideas to tangible projects. You need a touch of the creative writer to spell out detailed plans for three- to five-year periods.

Novice proposal writers sometimes come to the task convinced that clever people do not give away secrets until they have won the contract. But a good proposal must describe a project in enough detail to convince reviewers that they are learning what will happen at every stage of the project. Bidders to the U.S. Air Force, for example, are instructed to be specific: "The proposal shall not merely offer to conduct an investigation in accordance with the technical Statement of Work, but shall outline the actual investigation proposed as specifically as possible." Saying "The work described in the Request for Proposal will be performed as specified" is not much different from saying "Send money."

Of course, a proposal may include a technical design or a management plan that the bidder does not want disclosed. In that case, a Restriction on Disclosure notice, stating that information may not be disclosed for any purpose other than to evaluate the proposal, can be printed on the title page, and every sheet of data that is also so restricted can be marked: "Use or disclosure of proposal data is subject to the restriction on the title page of this proposal."

Strategic Planning for Funding Success

Solicited or Unsolicited?

Possibly the most important factor influencing your decision about the kind of response you prepare is whether the work you propose has been explicitly solicited by a sponsor. Proposals are said to be solicited when a sponsor formally announces that funding is available to conduct research in a specific area. Such an announcement may be called a request for proposal (RFP), a request for applications (RFA), or a notice of program interest (NPI). Proposals are considered to be unsolicited when they are submitted to an agency that has not circulated a formal request for research. You may also be confronted with a hybrid form: in these situations, the sponsor provides explicit proposal preparation guidelines, but research topics are not specified.

These differences affect the strategy you apply to persuade reviewers to support your project. Solicited proposals must address a problem in an area defined by the sponsor. They will be judged by the writers' ability to meet a specified need, to economize, and to deliver a quality product. These proposals may be measured against their competition on the basis of originality and importance within a discipline. Completely unsolicited proposals present the most severe writing challenge. Here you have the twin tasks of persuading potential sponsors that a problem or a need exists and persuading them that yours is the right group to solve the problem or meet the need (Figure 11.1a, b).

Enter the Right Competitions

Because proposal writing is absorbing and often exhausting, you need to enter the right competitions. If a request for a proposal is available, study it carefully. It will provide detailed information about the work requested and the document required to support your petition. A proposal has the highest chance of success when it is well matched to an assessment of the sponsor's needs. When an agency has explicitly listed

Desalination of Salt Water Using Wind Energy

An Unsolicited Research Proposal

M.S. Manalis
Environmental Studies Program
University of California
Santa Barbara, California 93106

March 15, 2006

Table of Contents

i

Abstract

The goal of this proposal is to carry out a feasibility study of coupling wind energy with desalination from small to large scale applications. In many developing and industrialized nations, food production is drastically inhibited by shortages of fresh water. By focusing on use of wind energy to carry out desalination of salt water, a better ratio of on-peak to off-peak electricity generated from wind turbines may be achieved, thereby increasing revenues from wind farms and providing fresh water for drinking, industrial processes, and possible use in agriculture. Important salt water sources include brackish wells, beach wells, agricultural drainage, and the ocean. Recent investigations reported in the literature indicate that wind driven desalination may show promise. California has been selected for study of wind driven desalination because over three quarters of the world's wind-turbine-generated electricity is produced in that state. In many parts of California (some along the sea coast), the wind power density is over 500 watts per square meter. A 34 megawatt state-of-the-art wind farm with a capacity factor of 27% operating in such areas will yield about 5000 acre feet of fresh water annually from desalination of ocean water. It is anticipated that the results will have application for developing and industrialized nations throughout the globe.

ii

Introduction

The developed countries are consuming fresh water at an ever increasing rate. America consumes about 360 billion gallons of potable water each day (Warfel et al., 1988), approximately three times the rate of consumption 30 years ago. This increase has resulted in the severe depletion of aquifers in the midwestern and southeastern United States.

The purpose of this investigation is to determine the feasibility of utilizing wind energy to desalinate salt water from brackish wells, beach wells, agricultural drainages, and the ocean.For the reasons given below, California will be the focus of the case study. At present 17,000 wind turbines are producing about 1% of California's electricity. Unfortunately, about seventy-five percent of this electricity is produced off-peak. This is a particularly serious problem because peak demand for electricity in California is increasing in relation to base demand. This jeopardizes the future value of electricity generated from wind farms.

Electricity is a perishable commodity; it must either be used or stored. Storage of electricity is expensive. When wind energy is used to desalinate water, the final product is not electricity, but potable water. It is much easier and less expensive to store water than electricity. Thus, when wind turbines are used to desalinate water, it is as if the wind turbines were operating on-peak one hundred percent of the time, a clear advantage over grid operating wind turbines.

1

(a)

Figure 11.1
The author of this research proposal uses the abstract and introduction to make a strong argument for the significance of the work he intends to do. He develops a clear list of objectives (a) and creates a graphic project schedule (b) to represent workplans. (Excerpted with permission from Dr. M. Manalis.)

Related Studies

Studies coupling wind energy and desalination have recently been conducted in Europe and America. In Spain, attention was directed toward providing drinking water by wind assisted reverse osmosis (Carta González et al., 1990). Feasibility studies of wind driven reverse osmosis led to the conclusion "that wind energy conversion systems may be an appropriate method of powering a reverse osmosis desalination system" (Warfel et al., 1988; Peterson, 1981). These studies indicate that when energy costs are high, wind driven reverse osmosis for desalination probably will be feasible. Cadwallader et al. (1977) reported that coupling of wind energy turbines to desalination equipment offers technological advantages which could be used to optimize the desalination process.

Recent research data in California regarding wind energy and desalination could not be located. However, water planners in Santa Barbara County considered using wind energy on Point Arguello for Desalination of sea water (Stubchaer, 1990). Documentation of this effort was not available. Manalis (1979) investigated a wind-driven vapor compressor for use in desalination of ocean water.

2

Objectives

This study will encompass the following broad objectives:

1. Analyze interface technology between desalination equipment and wind turbines.

2. Estimate economic and technological costs of using off-peak wind farm electrical output to desalinate sea water.

3. Evaluate the advantages and disadvantages of using wind energy on site to desalinate salt water.

4. Assess electricity-producing wind turbines as pollution offsets for desalination by conventional energy.

5. Describe the use of wind energy desalination plants to recharge ground water aquifers.

6. Determine economic and engineering feasibility.

7. Recommend sites for pilot plant if conclusions of study warrant such a recommendation.

Project Schedule

This study will be conducted over a twelve month period starting July 1, 2006, and concluding June 30, 2007, according to the following schedule:

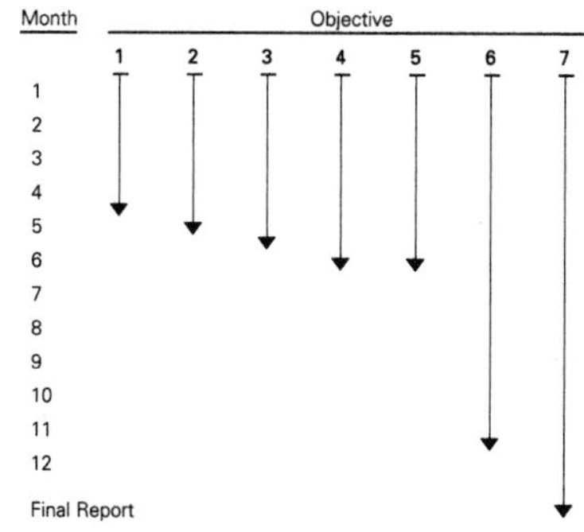

Methods

A case study approach will be used with primary emphasis on assessing the potential for developing wind energy desalination in California. The study will focus on California....

4

Conclusions

Preliminary calculations demonstrate that wind energy has the potential to produce significant amounts of fresh water. The city of Santa Barbara is proceeding with plans to build a desalination plant that uses conventional energy to produce 5000 acre feet of fresh water annually. This amount of water could be produced on Point Arguello by either one wind turbine wall array desalinating brackish water or four arrays desalinating ocean water.

References

Cadwallader, E.A., W.R. Williamson and J.E. Westberg (1977). The application of wind energy systems to desalination. U.S. Department of Commerce, Washington D.C.

Carta Gonzalez, J.A. and R. Calero Perez (1990). Optimization technical-economic of the desalination system worked by wind energy. European Community Wind Energy Conference, Madrid, Spain.

Manalis, M.S. (1979). Marine sciences and ocian policy symposium. University of California, Santa Barbara, p. 294.

Peterson, G. (1981). Wind and solar powered reverse osmosis units-design, startup, operating experience. *Desalination* 39: 125.

Stubchaer, James (June, 1990). Private communication. Water Management Consultant. Santa Barbara, CA.

Warfel, C.G., J.F. Manwell, and J.G. McGowan (1988). Techno-economic study of autonomous wind driven reverse osmosis desalination systems. *Solar and Wind Technology* 5: 549–561.

5

(b)

Figure 11.1 (continued)

areas in which it wishes to sponsor research, research on alternative topics will probably not be funded. The review process is likely to give the largest number of points to projects that are responsive to the agency's request. Proposal documents are not usually the right vehicles for arguing that an agency should be supporting research in extragalactic astronomy if the topics listed are all about geophysics.

Occasionally, proposers are encouraged to include alternative plans for satisfying a sponsor's request. In less restrictive cases, an RFP may say that for equal or even preferred consideration a prospective contractor is not limited to the suggested approaches but that any deviations must be fully substantiated. In other cases—and these are real challenges—you need to comply precisely with the terms and conditions stated in the RFP and also to submit a separate alternative proposal, along with a rationale indicating why the acceptance of the alternative proposal would be more advantageous to the sponsor.

Think in Two Time Frames

In the challenging work of proposal preparation, you need to think in two time frames: the time you need to prepare the proposal document and the time you need to complete the proposed research. In both cases, you match what you have to the sponsor's request.

For the first time frame, some competitions have no fixed deadlines, and some do. For example, most National Science Foundation (NSF) grants for research in education and engineering may be submitted at any time, though some NSF programs set target dates or deadlines for submissions to allow for their consideration by specially assembled review panels that meet periodically. If you know that you cannot get a strong proposal in on time, it may be best not to enter a competition.

For the second time frame, consider whether the research you are proposing is well timed for the announced term of the grant. Funding agencies like results. They will want to be convinced that the work you propose to do in 12 months is actually achievable in that period. Proposal reviewers are usually knowledgeable, and they will prefer funding a summer research project that can be completed in two months to funding a summer project that obviously will take two years. A graduate student requesting one month of research support to review the literature on solid, liquid, and hybrid rocket motors will appear to have a good sense of the work to be completed and the length of the task. An application for a full year of support to do such a literature review will not be very persuasive. Research proposals with long time frames present particular challenges. They need powerful evidence that the work you propose both requires (for example) 36 months of effort and can be completed in 36 months.

Read RFPs carefully, looking for information about preferred time frames. Improve your chances of success by proposing projects that fit the agency's guidelines about time. Consider, too, that many agencies acknowledge that research plans have natural phases or breakpoints, and they allow for proposing specific phases of projects. Progress reports are normally expected at the end of such phases, and follow-up funding for additional phases may be available.

Take Advantage of Help

Overcome any reluctance to take full advantage of assistance. Many agencies encourage you to contact program personnel before preparing your proposal. A meeting will help potential bidders determine if preparation of a formal submission is appropriate. Even if the agency discourages you from proceeding, the feedback you receive may help you develop subsequent proposals. Talk to colleagues who have dealt with the agency or sponsor in the past. Review funded proposals. This is not a do-it-yourself task. Successful proposals require negotiation: you have an idea; you call an agency to discuss your idea; you revise your idea. With every move you narrow the gap between what the agency wants and what you have to offer.

If you are entering a new research area, conduct a literature search on the topic to get a better grasp of any published findings, relevant methodologies, and possible collaborators or competitors. Be sure to consider any political significance of your project. If your research bears on the domain of another scientist or engineer, consider ways to diffuse potential clashes. You might, in some cases, frankly disclose your interests and involve the other researcher at the planning stage.

Use Evaluation Criteria as Planning Tools

When a sponsor provides criteria for evaluation of proposals, study them carefully at the planning, drafting, and editing stages. To deal effectively

Proposal Evaluation Criteria

U.S. Department of Transportation
Annual Solicitation for University Researchers

- Merit of the Technical Approach.............(possible **40** points)
- Merit of the Management Approach......(possible **30** points)
- Qualifications of Proposers.....................(possible **20** points)
- University Support..................................(possible **10** points)

Figure 11.2
In the weighted criteria announced by the U.S. Department of Transportation, a strong management approach could compensate for insufficient university support.

with reviewers, you must continually consider their constraints and requirements. If criteria for awards are not published in the RFP, it is sometimes possible to get more information by telephoning the funding organization.

Some agencies list areas of evaluation without announcing the weight given to each one. Other agencies provide breakdowns of review categories, so that you can consider the sponsor's weighting as you decide whether to bid. Bidders to the U.S. Department of Transportation, for example, learn that an unusually sound management approach could offset the problem of insufficient university support (Figure 11.2).

Learn about the Review Process

Proposal writers have an important factor in their favor: agencies and other sponsors need good project proposals. When research funding is available, referee teams hope to award their support to someone. Referees want, most of all, to be vindicated in their choices. They have a stake in awarding funds to promising projects. They need more than good concepts; they need evidence that you can meet claims and deadlines. To get well-conceived and well-written documents, agencies often provide extensive feedback to proposal writers.

Proposal funding is, however, not entirely rational. The effectiveness of the peer review system for evaluating research proposals is a matter of vigorous debate. Critics of peer reviewing argue that reviewers are biased in favor of proposers at the more prestigious universities and that real conflicts of interest can arise between reviewers and applicants. In this view, the system encourages researchers to write devious proposals, not discussing innovative ideas for fear that disclosure will tip off competitors who review the proposal. Critics of the system regard the protective cover of "blind reviewing" as a myth. They point out that knowledgeable scientists in a given field can often deduce the name of the applicant from the very subject of the proposal—and even if they can't, references indicate the applicant's past research and published papers.

Learn as much as you can about the review process for the proposals you submit. The National Science Foundation (NSF) announces that all proposals are reviewed by a scientist, engineer, or science educator serving as an NSF program officer, and usually by three to ten other experts in the field represented by the proposal. Proposers to the NSF are invited to suggest names of persons they believe are especially well qualified to review their proposal and also, giving reasons or not, to suggest persons they would prefer not review the proposal. Most agencies advise that proposers allow at least six months for programmatic review and processing.

Get Approvals in Advance

Proposals that commit the resources of an institution must be approved by an appropriate administrator before they can be submitted. Many universities have administrative units whose responsibility is the administration of contracts and grants. These offices can provide useful guidance. External approvals may be needed as well: A team of physicists at the University of California, Santa Barbara, cannot, for example, guarantee that research will be carried out at Argonne National Laboratories unless they have obtained permission to do so.

In industrial settings, approvals are equally important. An engineering group employed in an electronics firm cannot submit an external proposal without management consent. A proposal is a legal document as well as a sales and a technical document: It binds a company to the statements it contains, and it must be signed by someone authorized to make such promises.

Systematic Proposal Preparation

The proposal is a written product that sets forth the design of a technical product. These tasks mutually govern each other. The tasks of writing the proposal and those of doing the work are often analogous. Both require a systematic approach. Both require knowledge of logical work units. Both require careful estimation of completion time. Both require allocation of responsibility. You can expect, therefore, that the same project management tools used to monitor the progress of writing the proposal are used later to monitor the work defined in the proposal.

Typically, preparing the research plan and preparing the document are not neatly isolated stages. Many sections of proposals are written before calculations have been completed, and tidy schemes for proposal preparation and systematic editing seem almost comic to proposal writers working against deadlines. The following preparation model applies to many kinds of proposals, from relatively brief unsolicited documents written by one researcher to multivolume solicited documents written by a team of 20 or more and coordinated by a proposal manager. Naturally, not every step will apply to every proposal you write, and you will not always have time to attend to every step. Still, the more you are requesting, the more attention you should devote to preparing an outstanding document.

Study the Request for Proposal

The most important criteria for proposal writing are the explicit instructions in the RFP. Follow proposal preparation instructions exactly. You must provide what the RFP asks for. You do not want your work to be rejected because it has exceeded explicitly stated page limits or does not contain a budget statement matching provided instructions. Read the RFP more than once—and if your proposal will involve other researchers or writers, be sure that all members of the group read the entire RFP, not just their own sections.

The instructions for proposal preparation may include a number of forms to fill out and submit with your document: cover sheets, budget sheets, forms for biographical sketches, checklists, mailing labels, and more. Be sure to use each applicable form, following preparation instructions exactly.

Whatever the proposal, it will usually have a prefatory section with a letter of transmittal, cover page, table of contents, list of figures and tables, and summary; a main section with technical, management, and cost details; and some appended items. Complex industrial proposals are often produced as separate technical, management, and cost volumes. Each volume is then evaluated by a specialist. In preparing your document, take care to comply with requirements and also to make your compliance visible with sections, titles, and headings that match instructions in the RFP.

If no instructions for proposal preparation are provided, you may select elements with more freedom than writers who must comply strictly with specified requirements. The most sensible plan, however, is to prepare proposals in conventional ways, answering the standard questions. The proposal format displayed in (Figure 11.3) is widely used, covering technical, management, and cost areas.

Turn Requirements into Outline and Compliance Matrix

Use the instructions in the RFP to draft an outline for each section of the proposal, following the exact order and numbering system in the RFP. When a sponsor provides proposal preparation instructions, a good practice is to prepare a compliance matrix, whether one is requested or not, to indicate your adherence to specifications and to show evaluators where they can find your response to each item required in the RFP. Compliance matrixes (also known as response indexes) indicate the correspondence between the required sections and your document (Figure 11.4). Because they can be used to provide a graphic view of what still needs to be done, compliance matrixes are also useful as task management tools.

Make Plans for Time Management

Leave time for your proposal efforts. Industrial bidders to the U.S. government read the government publication *Commerce Business Daily*

Proposal Format

Front Matter

- Letter of Transmittal
- Cover Page
- Project Summary
- Table of Contents
- List of Figures and Tables
- Compliance Matrix

Body of Proposal

- Technical Section
- Management Section
- Budget

Appendixes

- References
- Curriculum Vitae
- Supporting Details

Figure 11.3
Standard format for a formal research proposal. Most proposals explain what work will be done, who will do it and why they are qualified, where they will do it, and in what time frame and location and at what cost they will carry out the plan.

COMPLIANCE MATRIX

REQUIRED	FULLFILLED PAGE(S)
FRONT MATTER	
Letter of Transmittal	
Cover Page	
Abstract	ii
Table of Contents	iii - iv
Liste of Illustrations	v - vi
Compliance Matrix	vii
TECHNICAL SECTION	
Research Objectives	1 - 3
Methodology	4 - 6
Anticipated Results	7
MANAGEMENT SECTION	
Task Breakdown	8 - 9
Timetable	10
Qualifications of Participants	11 - 14
Related Experience	15 - 16
Facilities	17 - 18
BUDGET SECTION	
Cost	19 - 20
Justification	21 - 24

Figure 11.4
A compliance matrix shows proposal evaluators that the document responds to requirements in the request for proposal (RFP).

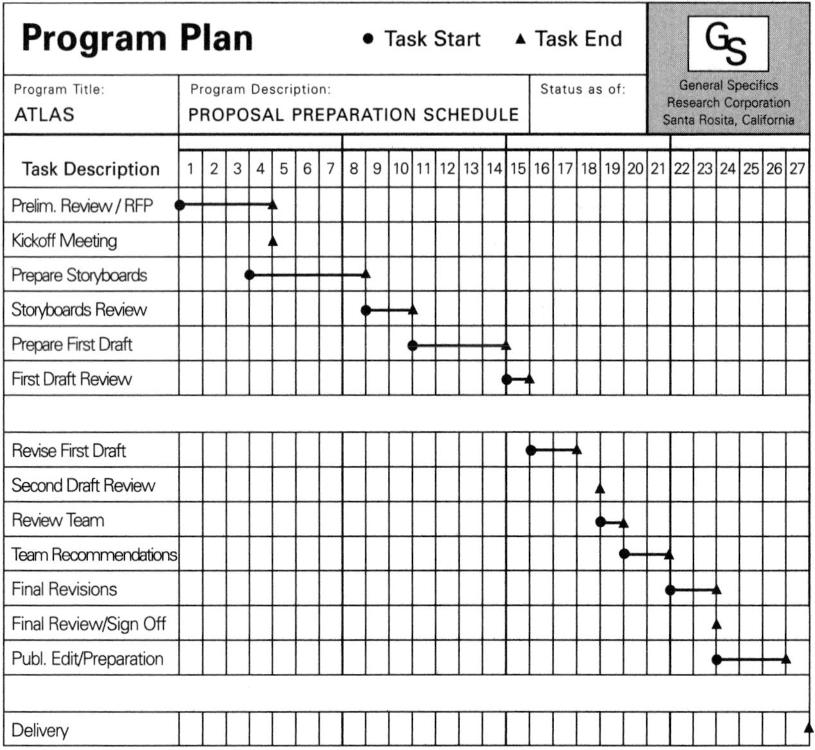
Program Plan
● Task Start
▲ Task End
GS
General Specifics
Research Corporation
Santa Rosita, California
Program Title:
ATLAS
Program Description:
PROPOSAL PREPARATION SCHEDULE
Status as of:
Task Description
1 2 3 4 5 6 7 8 9 10 11 12 13 14 15 16 17 18 19 20 21 22 23 24 25 26 27
Prelim. Review / RFP
Kickoff Meeting
Prepare Storyboards
Storyboards Review
Prepare First Draft
First Draft Review
Revise First Draft
Second Draft Review
Review Team
Team Recommendations
Final Revisions
Final Review/Sign Off
Publ. Edit/Preparation
Delivery

(a)

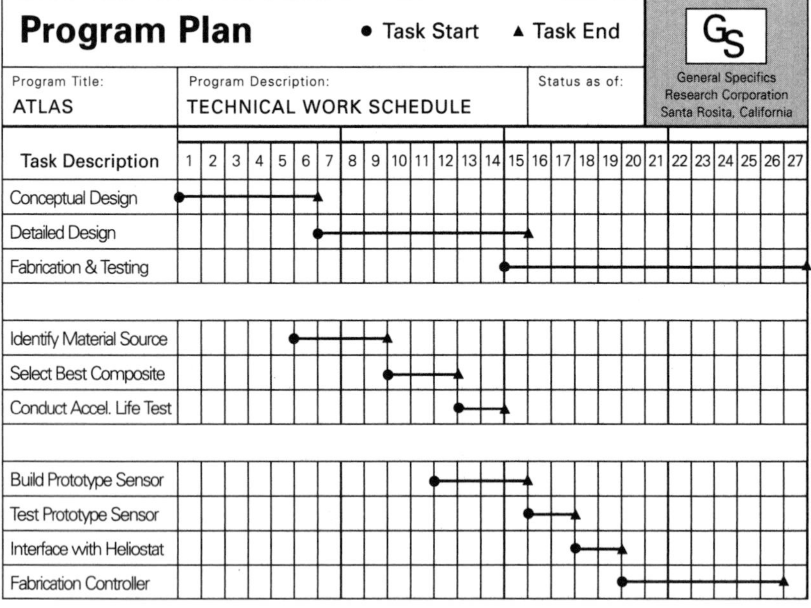
Program Plan
● Task Start
▲ Task End
GS
General Specifics
Research Corporation
Santa Rosita, California
Program Title:
ATLAS
Program Description:
TECHNICAL WORK SCHEDULE
Status as of:
Task Description
1 2 3 4 5 6 7 8 9 10 11 12 13 14 15 16 17 18 19 20 21 22 23 24 25 26 27
Conceptual Design
Detailed Design
Fabrication & Testing
Identify Material Source
Select Best Composite
Conduct Accel. Life Test
Build Prototype Sensor
Test Prototype Sensor
Interface with Heliostat
Fabrication Controller

(b)

(⟨http://cbdnet.access.gpo.gov/⟩) for notice of upcoming solicitations. In a reasonably typical scenario, advance notice of a solicitation will appear on March 6, the RFP will reach the potential bidder on September 25, and the completed proposal will be due on November 1. Academic proposal writers receive mailings from potential sponsors, so they can plan proposal responses many months in advance.

Nevertheless, proposals are usually written under pressure, when you have technical work as well as proposal writing to do. Often in a matter of one or two weeks, a project concept must be refined, a team assembled, and a detailed document prepared. This complex process must be thoughtfully sequenced and coordinated to prevent a waste of resources. It must include steps that the proposal group may actually enjoy doing, like technical brainstorming, and steps that many group members will resent, like leaving valuable time for proofing, printing, binding, and delivery.

When you work out a routine for proposal writing, you need to allocate time for each step by first identifying the submission deadline and then backing up to the present (Figure 11.5a, b). Many proposal writers use project management software to track their progress. The best-made schedules will change. You may need more time than you had anticipated to prepare a budget section and considerably less time to prepare a list of related contracts.

Allocate Team Responsibilities

Managing a group preparing a proposal can be as challenging as managing the research itself. In academic settings, the group writing the proposal will probably be the same one slated to carry out the work. Most research universities have offices of contracts and grants to give advice at various stages. In industrial settings, many more people are part of proposal preparation. In addition to the research group, proposal managers, budget analysts, technical managers, artists, and technical writers are

Figure 11.5
Many proposal preparation teams plan and track their tasks with Gantt charts. This group has planned twenty-three days for writing and three days for producing their proposal (a). Note, however, that the technical work that is the subject of the proposal goes on at the same time (b). Few engineers or scientists are able to devote full time to proposal writing!

often involved. The team's ability to collaborate is a decisive factor in the success of both the proposal document and the funded work.

We think that groups work best when they meet often and when their assignments and responsibilities are visible and explicit. One person should agree to be proposal manager. Every group member should read the entire RFP. Every member should receive an annotated proposal outline with specific allocation of responsibility. Every member should know who is responsible for each part of the proposal. Annotated calendars, printouts of graphics charting project progress, and the compliance matrix should be displayed in prominent places. The group should establish regular meeting schedules, and members should receive explicit instructions about preferred format, writing, and design strategies.

In some industrial settings, the storyboarding method is used to manage collaboration. The proposal manager prepares an outline to match requirements in the RFP, and each member of the writing group takes specific portions of the outline. Authors receive preprinted forms, each representing a two-page spread in the final proposal. They fill in the left side of the storyboard with a thesis sentence and notes about the point to be made in response to their section of the outline. They fill in the right side of the form with rough drawings of illustrations to support the point as well as captions for the illustrations (Figure 11.6). After pinning their storyboards to the wall of a large room, team members can review the document as they walk and talk their way around the room (Figure 11.7).

Storyboarding is helpful because it facilitates revision. Each two-page module can easily be improved without changes to the rest of the document. The process also facilitates review as it makes inconsistencies obvious. It coerces writing that is responsive to the requirements of an RFP, and it makes effort (or lack of effort) visible: Blank spaces will show where a delinquent engineer's storyboards should be. In addition, storyboarding produces an efficient document design: Tables and figures are always located on the right facing page, directly across from the text passage in which they are discussed. Yet some proposal writers find the method overstructured. To be successful, storyboarding requires a firm commitment from a proposal manager, because it never just happens.

STORYBOARD

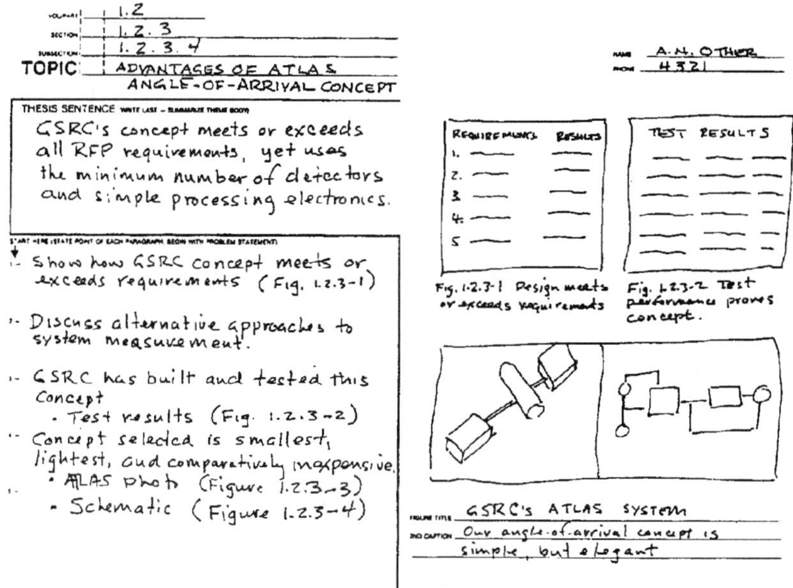

Figure 11.6
This storyboard is a draft of text (left) and graphics (right) for two pages of the Atlas proposal. (Courtesy Hugh Marsh.)

Prepare Style and Format Guides

The most efficient way to achieve consistency in proposals is to prepare style and format specifications. Style guides may be as informal as a single-page handout asking all writers to do three things: (1) use the active voice, (2) put important ideas in the first sentences of paragraphs, and (3) use a hyphen when "strip-mine" is a verb. Or they may be lengthy manuals covering numerous issues such as how to prepare mathematical material; preferred spellings, abbreviations, and acronyms; and grammar, capitalization, and hyphenation. The alternative to establishing style guidelines in advance is to establish and apply them at the end, when editing time is usually better spent imposing consistency in mechanical matters such as renumbering equations.

Consistency in format is at least as important as consistency in style, showing proposal evaluators that you have prepared the document with

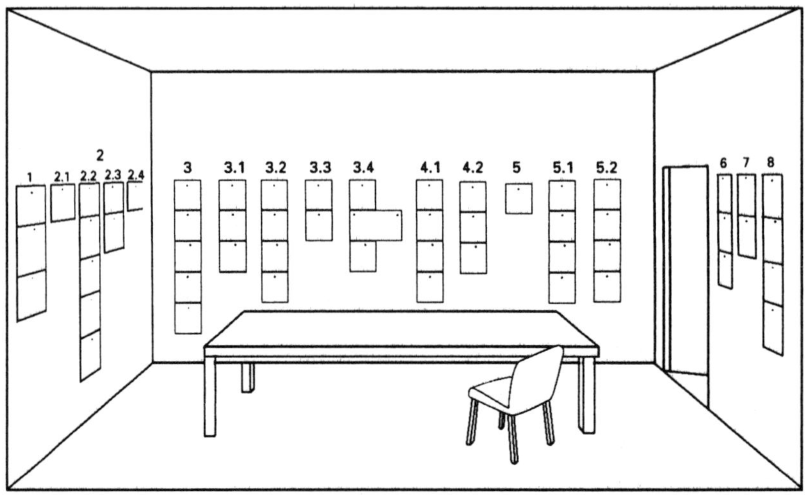

Figure 11.7
Proposal team members pin their storyboards to the wall, and the entire document is reviewed *before* the final draft stage. (Courtesy Hugh Marsh.)

care. Your team may find it helpful to distribute samples of finished pages, with heading styles and sizes highlighted, and samples of completed illustrations. Simple format instructions such as "Use Courier font in 10 point, design all figures to fit either one-half page or a full page, and design all figures to be read vertically" may be all you need to ensure that the proposal looks carefully produced.

Format specifications can be stored electronically. You can create templates of basic pages with predefined styles of headings, type sizes, fonts, margins, spacings, indentations, and other features. Routines that manage numbering systems for elements like headings, references, and equations are widely available.

Facilitate Electronic Submission

Many funding agencies now require electronic submission of proposals. The National Science Foundation paperless proposal and award initiative is called FastLane, an interactive real-time system used to conduct NSF business over the Internet (⟨https://www.fastlane.nsf.gov/fastlane.htm⟩). Principal Investigators (PIs) at registered FastLane institutions prepare their proposals on-line, and authorized co-PIs can access and modify the proposal. Access is also granted to the proposers' Sponsored Research Office for comment and approvals. An electronic proposal is not processed, however, until the cover sheet and certification page are printed, signed, and mailed to NSF.

Proposal Content

Front Matter

Letter of Transmittal A letter of transmittal (or a memo, in the case of an internal document) should always accompany your proposal. As shown in Figure 11.8, the letter should identify the solicitation you are responding to and give a brief overview of proposal contents.

Cover If a preprinted cover sheet is supplied in the RFP, be sure to use it. If you need to design your own cover, provide the project title and name the proposing organization, the potential sponsor, the date, and program solicitation number (Figure 11.9). A proposal title should be brief, informative, and intelligible to a scientifically literate but nonspecialist reader. Take advantage of the visibility and prominence of the title to teach reviewers about your idea and to sell its advantages.

Project Summary In a project summary, briefly describe the problem addressed in the study, the methods used, and the expected results. Summaries are typically one to three pages. Think of the summary as a freestanding document, one that may actually have much wider circulation than the rest of the proposal. Be sure that readers can profit from the summary without reading the main body of the proposal: Do not refer to tables or figures that appear elsewhere; define acronyms and avoid abbreviations. Some RFPs ask that the summary be written at a level appropriate for an audience of educated but nonspecialist readers. In these cases, summaries of successful proposals may be used in agency reports and news releases.

Some RFPs ask for both summary and abstract. Like a summary, an abstract may have a life of its own and be read by far more readers than those who evaluate the proposal. Abstracts are typically briefer than summaries (one paragraph of approximately 150 words), and they are

CGK Engineering

15 December 2004

Dr. C.W. Low
Director of Research
California Electric Company
Santa Rosita, CA 93109

Dear Dr. Low:

Enclosed is our proposal for a Tidal Mini-Hydro Power Plant feasibility study. The proposal was written in response to your Solicitation #AN248, and we have prepared the document in exact compliance with instructions provided in the RFP.

Tidal mini-hydro power plants can become an important supplemental source of energy for many specialized locations. Because of their potential to alleviate future fossil fuel shortages, a feasibility study of site selection criteria and technical requirements is both timely and worthwhile

We look forward to answering questions you may have and to working with you.

Yours truly,

Brian Kato

Brian Kato
Tidal Mini-Hydro Division

Enclosures

2535 West Armadillo • Suite 205 Santa Rosita • California 93016

Figure 11.8
A letter of transmittal should identify the competition you are entering and provide a capsule version of proposal contents.

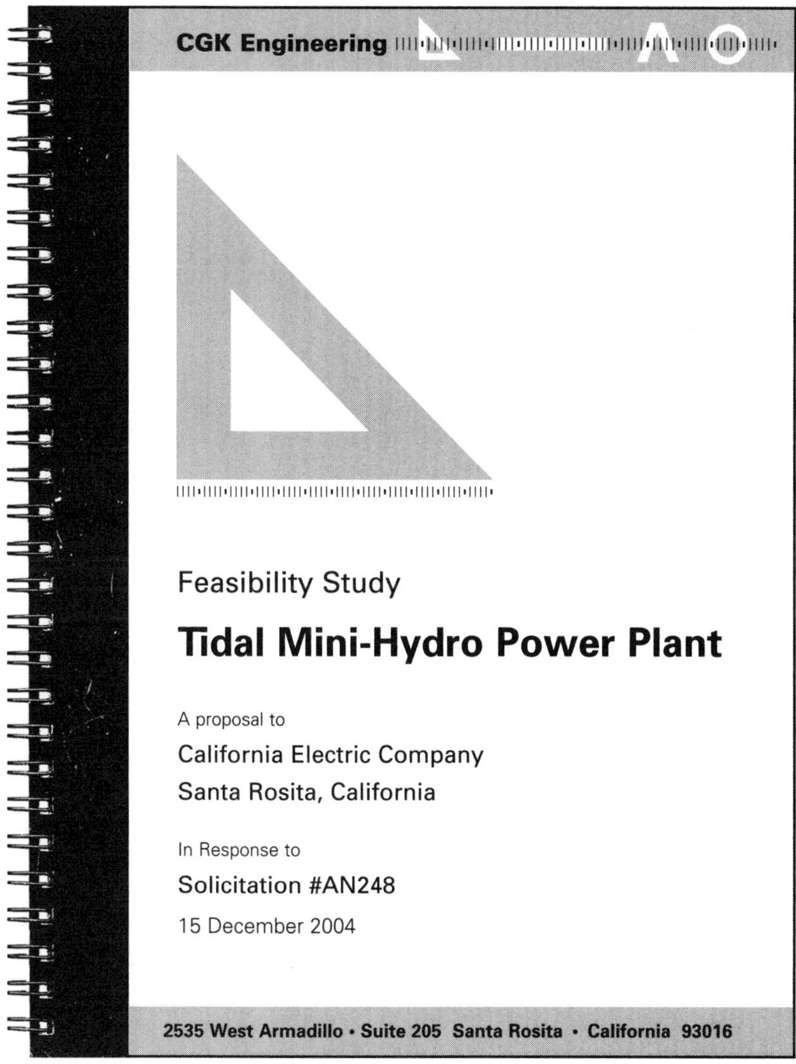

Figure 11.9
The cover page is the most prominent element in a proposal document. An informative title and attractive design can help to persuade reviewers of the value of your idea.

written for the same specialist readers who will read the proposal (see discussions of abstracts in Chapters 13 and 14).

Table of Contents The table of contents serves as an organizational map of your proposal, helping evaluators locate relevant material. In the table of contents, list section headings and name the elements contained in appendixes. Provide a page number for each element. Many proposal writers provide two tables of contents: a brief version with first-level headings only and an expanded version with headings at second, third, and even fourth or fifth levels.

List of Figures and Tables List all figure and table titles and their page numbers. The list of figures and tables is highly visible and widely used by technical readers. As with headings, you can use titles to inform and persuade. Instead of writing perfunctory titles like "Filtering system," you can write titles that lead evaluators to the conclusion you hope they will reach: "Filtering system has been modified to exceed requirements."

Compliance Matrix Whether or not one is called for in the RFP, a compliance matrix (Figure 11.4) indicates that you have paid careful attention to the sponsor's requirements. It also tells reviewers where they can find your response to each required section.

Body of Proposal

Technical Section In the technical section, identify the problem and its significance, state the objectives of the proposed investigation, and provide a clear statement of the work to be undertaken. Outline your approach to the research, noting significant alternatives and your reasons for not pursuing them. In many cases, you are also expected to review earlier work and related studies; advice about preparing such reviews is provided in Chapter 14.

Management Section The management section names the personnel who will do the proposed work and the facilities in which the work will be done. It also contains highly detailed task breakdowns and work schedules. Management sections are the place to argue for the qualifications of the principal investigators and their associates. Relevant highlights of curricula vitae can be summarized, and lists of previous related contracts can be provided.

Budget In your budget, provide cost details for salary and benefits, and justify each number. Include indirect costs (overhead), as well as direct costs like materials, equipment, salaries, and travel. Many sponsors provide preprinted sheets for budget calculations; to comply with the RFP, you must use those sheets to record your request.

Appendixes

References List references to previous papers, documents, and discussions that have been used in preparing the proposal.

Supporting Details When appropriate, include items like curricula vitae, copies of publications of principal investigators in areas related to the proposal topic, lists of previous related contracts, letters of reference, and detailed and oversized figures and tables.

Business Plans

Engineers and scientists are increasingly interested in raising money to start their own technical enterprises. In such cases, they need to write a business plan, a specialized form of proposal pitched to an audience of investors or bank loan officers. A business plan has much in common with a standard proposal: It must demonstrate keen understanding of the project you wish to undertake as well as of your qualifications for accomplishing what you say you will. Like a formal proposal, a business plan is a detailed blueprint for the work you will do and the time in which you will do it. The major difference is in purpose. While the explicit goal of a research proposal is likely to be an enhanced understanding of some unsolved problem in science or engineering, the explicit goal of a business plan is to raise money for an enterprise that will produce a financial profit. For that reason, business plans require focused explanations of how your technical plan enables you to do something that others can't (or don't) do and that customers will be willing to pay for. You need to understand the market for your concept as well as the competition, and you need to provide solid evidence that you and other members of your team have the experience and credibility to use investors' money wisely.

Format for a business plan can vary, but the modules in Figure 11.10 are typical. Many business plan consultants advise authors to treat the executive summary as the most important section of the business plan. It will be more widely read than any other section and should contain a self-sufficient, well-reasoned case for providing financing for your project. In many cases, a face-to-face meeting with potential investors is also required. You need to be able to speak enthusiastically about the critical elements of your venture and to be prepared for highly critical questions.

Stressing the Strengths of Your Ideas

Because a proposal is a sales document, you need to be able to identify and emphasize the features and benefits of your idea. At the paragraph and sentence level, in section previews, in well-designed graphics, and in captions to illustrations, skillful proposal writers can reinforce powerful arguments for the value of their plan. Instead of writing a perfunctory caption like "Project Placement," you can write something that may help evaluators to reach positive conclusions about the plan described: "Project Placement Takes Maximum Advantage of the Expertise of Three Research Groups."

Keep in mind that proposals are likely to be reviewed by readers with widely different interests and levels of technical understanding. Because each reviewer will be interested in different aspects, plan to repeat key ideas and make sections of your proposals as nearly freestanding as possible. While full details of the work are provided in the technical volume, a budget analyst may be reading only the cost section. Be sure to provide at least a capsule of information about key aspects of your project in every volume of the proposal.

Give reviewers the impression that you are confident of success by using the present tense for general descriptions and the future tense for actions in the future. Write as though the funds and approvals have been granted. A proposal style dependent on conditional verb forms is awkward: 'If and when funding were to be granted, we would at the time of

Business Plan Format

Front Matter

- Letter of Transmittal
- Executive Summary
- Table of Contents
- List of Tables and Figures

Body of Business Plan

- Introduction
- Market Assessment
- Financial Plan
- Marketing Plan
- Company Structure
- Implementation Schedule

Appendixes

- Curriculum Vitae
- Letters of Support
- Credit reports

Figure 11.10
Standard format for a business plan. The executive summary may be more widely read than any other section.

the second phase of the project develop test equipment." Compare "In Phase 2 we will develop test equipment."

In some particularly competitive industrial settings, proposal groups create a "win theme matrix." They list strong thematic phrases or sentences that argue for the merits of their idea—and they write these phrases into every section of the proposal, using the matrix as a checklist to assure that they have repeated their win themes.

Preparing an Attractive Document Package

Though proposal evaluation schemes never award any points for visual attractiveness, the way a document looks can convey a powerful sense of your professionalism and competence. Even when a proposal is written in accord with rigid and challenging page limits, its design elements can facilitate navigation through the document. Tabbed section dividers and informative page headers help busy reviewers read your document efficiently. Judicious selection of type styles and sizes will signal what elements are more important than others. A heavyweight or even laminated proposal cover may assure that your document will hold up to multiple reviews. A logo, proposal title, and organizational name on every page will serve as a reminder to reviewers of who you are and what you are selling. Successful proposals are often exceptionally attractive documents, with numerous foldouts, photographs, and other artwork set on pages designed with great care.

But in selecting design options, be sure to read the RFP for guidelines. Many funding agencies provide explicit instructions for utilitarian design and packaging.

Your Proposal Writing Program

Resubmitting

It is highly unlikely that every proposal you write will be funded, and it is hard to know exactly why one proposal is successful and another is not. If your project is denied, you can still get some benefit from the work you have done by finding out why, a process called "debriefing." The amount of information you can receive about the deliberation process varies from agency to agency, although the identity of reviewers is never revealed.

If you have submitted your proposal to the National Science Foundation, you will automatically receive a debriefing. When the decision is made, the principal investigator receives copies of reviews (excluding the names of reviewers), summaries of review panel deliberations, a description of the process by which the proposal was reviewed, and details about the decision, such as number of proposals and awards. Other agencies will provide debriefings with a written request, usually within 30 days after the announcement of the final selections.

Depending on what you learn in your debriefing, you may want to resubmit the same proposal or one substantially like it in a future competition. Possibly it was technically excellent but did not fit the agency's research priorities for that year. If you identify weaknesses in your project proposal, consider ways of salvaging the concept. In some instances, you may focus the proposal on another area and apply to another agency.

Creating Document Databases

If you regularly write proposals, you will want to create computer files of standard information, text, and graphics. Many proposal sections—including curricula vitae, drug-free workplace plans, related experience, and management history—are likely to be required in nearly the same form for any project you may bid on. Instead of compiling and typing these chunks of standard text each time a proposal is created, you can record and save them as separate files that can be quickly tailored and inserted into new documents as needed (Figure 11.11).

You will, of course, want to review the old files as you use them, but they are easily updated to contain the most relevant text and graphics. An additional benefit is that these data files can be made available for use by other proposal writers in the organization.

Staying Informed

A proposal-writing program requires careful planning. You must know the needs of your discipline; you must have good information on funding sources and their requirements. Keep a file of agency announcements,

Typical Amber Engineering Proposal Structure

Section Number	Description	Text is Usually
—	Executive Summary	New
1	Introduction	New
2	Technical Approach	New
3	CMOS Technology	Standard
4	Test Approach	Modified
5	Related Experience	Standard
6	Program plan	Modified
7	Proprietary Technology	Standard
8	Statement of Work	Modified
9	Organization	Standard
10	CV's	Standard

Figure 11.11
Of 11 sections included in proposal by writers at Amber Engineering, Santa Barbara, CA, 5 are standard in all proposals, 3 are modified, and only 3 are new. Reusing stored text will yield improved accuracy in content and consistency in format. (Courtesy of Stan Laband.)

RFPs, and NPIs. Keep another file of new project ideas that occur to you or to members of your team. Articles from the literature may suggest new research possibilities for your field. To develop new ideas, go to conferences, check the *Annual Register of Grant Support* (R. R. Bowker) and keep up with the literature. Look for technological developments that make new research feasible; some researchers sift through patent literature for ideas. The Federal Research in Progress Database (FEDRIP) provides access to ongoing federally funded projects in physical and life sciences as well as engineering (⟨http://grc.ntis.gov/fedrip.htm⟩). The home pages of agencies like the National Science Foundation (⟨http://www.nsf.gov⟩), the National Institutes of Health (⟨http://www.nih.gov⟩), and the Department of Transportation (⟨http://www.dot.gov⟩) are good sources for proposal announcements and deadlines.

If you depend on the support of sponsored research, you must have a long-range funding strategy. If you are writing proposals in June for next winter's support, you're in trouble. Most research proposals take from four to nine months for review, and you may need to be thinking two to three years ahead. Track new project possibilities and funding sources. Keep agency application deadlines prominent on your work calendar, and meet each annual application deadline with one or more new proposals. A professional researcher may have five to ten proposals circulating at once. To write proposals on this scale, you obviously need to work out a detailed application routine.

Proposals as Part of a Continuum

Proposals are crucial documents in the production of scientific knowledge, providing access to the funding and approvals that make research possible. When proposals are successful, they create more communication tasks. They lead to writing projects such as progress reports, final reports, conference proceedings, and journal articles. Each document form disseminates information to wider and wider audiences.

Proposals also create occasions for speaking about your ideas. When you work on a proposal, be prepared to talk persuasively about the project, on occasions ranging from informal telephone conversations and hallway meetings to formal presentations with potential sponsors. In active professional settings, you should always be primed to talk about your research and argue for your position.

12

Progress Reports

Audiences
Formats and Schedules
Organization
Design and Distribution
Document Databases

■

You're in a bind. Your third progress report is due tomorrow, and the news is bad. Less than halfway into a 12-month project, you've fallen behind schedule and used close to half your budget. You still hope to compensate for lost time and extra expenses, but now you must tell your manager about these difficulties. Unfortunately, you've already compounded your problem. Last month, you avoided reporting the project's deficiencies, hoping that you could solve the problems quickly. Your manager will be unpleasantly surprised that your last report was overly optimistic.

Progress reports are always intermediary, never final documents. They track, evaluate, and archive the work of science and engineering. Progress reports are bridges, spanning the time between the beginning and the end of a project—from the projections in the proposal to the reality of the work.

In the proposal that initiates a project, you describe tasks and activities that have not yet happened as though you know what will happen at every stage of the work. You estimate, guess, project. Your proposal has minimized ambiguity. Instead, you have described a predictable series of events leading to a successful outcome.

In contrast, you write a progress report as you do the work. Your report might tell the same story as the proposal, or it might tell a different one. A proposal says, "This is what will happen"; a progress report says, "This is what actually happened." The baseline against which progress is measured is what you said would happen: at what cost, in what time frame, with what researchers involved, and with what deliverables delivered.

Progress reports may be formal documents written to satisfy external funding sources, or they may be informal documents that track and monitor employee activity within an organization. In many industrial settings the progress report (sometimes called a status report, a morning report, a briefing, or a weekly) is one of the most frequent and routine pieces of writing for technical professionals. Typically, each researcher prepares an account of activities for a specified period and passes the account upward to a manager. The manager writes a summary progress report for an entire department and passes this more comprehensive account upward, to a manager at a higher level. At the highest level, a manager will be able to report on progress for numerous projects.

Audiences

Readers of progress reports are likely to be knowledgeable about the technical areas you describe and also deeply concerned with the status of activities. They want to know what has been done and what needs to be done, what problems you have encountered, and how likely you are to stay within a previously agreed-upon budget and schedule. They want to know if you are spending their money and your time in ways that will yield desired outcomes.

Because such readers are typically worried about the project, writers sometimes feel constrained to use the progress report to allay anxieties rather than record problems. But the progress report is the vehicle for assessment, evaluation, and possibly for renegotiation. Your report may cause the project to take a new direction, with revised goals. You might be tempted to use the progress report for claiming credit for your achievements or crowing about the excellent match between your projections and what has actually happened. Don't. It is safer to report achievements modestly and to report problems as soon as possible. Progress reports have been entered into legal proceedings. You are on solid ground if you record what has happened, not what you want your sponsor to think has happened. Recognize that authors of progress reports are not always able to record desirable advances toward the orderly achievement of a goal.

Your progress report should address the concerns of your readers by including a section devoted to assessment and evaluation. A genuinely useful progress report will include interpretation of results as well as raw numbers.

Formats and Schedules

In writing a proposal, you follow instructions provided in the request for proposal (RFP), or you create a responsive format of your own. In the same way, a progress report follows instructions provided by your sponsor or your supervisor, using the forms and formats specified, or you create your own functional document. For many government contracts, the frequency of progress reports will be specified in a contract data requirements list (CDRL), with details of form, length, and style provided in a data item description (DID).

Progress reports vary widely in form and length, from a one-page memo, letter, or preprinted form to a several-hundred-page bound volume with a separately bound appendix. In tracking a multiyear project, progress reports are often long at some stages, perhaps annually, and very brief at others.

Progress reports tend to be less formal than proposals or research reports. Each one summarizes progress since the preceding report, and the earlier report is always summarized, subsumed, and superseded by a new one. Progress reports have some elements in common with proposals and formal reports, but they are always distinguished by their focus on

Project status
Measurement of achievements against projections
Problems encountered
Work completed

Work remaining
Evaluation

Figure 12.1 lists possible elements in a progress report. In selecting elements, remember that all progress reports in a series should have the same format. For example, if you have included a section called "Problems Encountered" in the first report, you should include the section in the second report, even if all you have to say is that you have encountered no new problems.

The schedule for submitting progress reports is established in a proposal (Figure 12.2) or by organizational practice. Reports may be required weekly, monthly, bimonthly, annually, or on some other schedule. Often, one kind of report is required weekly while another kind is required semiannually. Whatever the schedule, you must follow it strictly, even if your work has not gone as you had planned.

Organization

Readers review progress reports to get answers to their questions about project status: perhaps monitoring employee productivity, perhaps keeping an eye on costs. A well-designed progress report will be modular, with modules clearly labeled, so that readers can skip over sections they do not need to read and proceed quickly to sections that interest them. No reader will want to spend time hunting for information. Use headings for each new topic area, and provide a table of contents to enable skimming. Consider using labeled tabs to divide and mark sections in longer reports.

Figure 12.3 illustrates alternative ways to present information in a progress report. Progress report 1, the more conventional of the two, is organized chronologically, with the emphasis on the current state of the project. In this version, the tasks are subordinated. Progress report 2 is organized by task, and in this version the current state of the project is subordinated. If a sponsor has not given explicit instructions for preparing a progress report, you can choose a presentation style that best reflects the work. Writing style in the narrative sections of progress reports is sometimes relatively formal, resembling the style of a proposal or final report, and sometimes informal, with phrases acceptable in place of complete sentences.

Progress Report Format

Front Matter

- Letter of Transmittal
- Cover Page with Date and Number in Series
- Table of Contents
- List of Tables and Figures
- Project Summary

Body of Progress Report

- Work Accomplished to Date
- Plans for Next Reporting Period
- Problems Encountered
- Appraisal of Progress to Date
- Recommendations

Appendixes

- Schedule
- Supporting Details

Figure 12.1
Possible elements in a formal progress report. If you need to prepare a series of progress reports for a single project, use the same elements and formats for each.

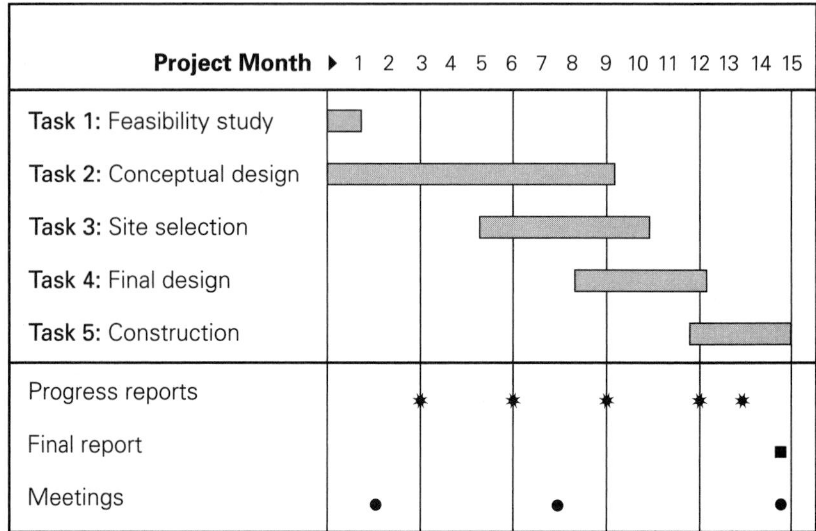

Figure 12.2
Five progress reports are scheduled during the 15-month period of this project. The schedule was established in the proposal.

Progress Report 1

Work accomplished to date:
- Task One: Evaluate Site
- Task Two: Treat Existing Coal Pile
- Task Three: Install Storage Pile Sprinklers
- Task Four: Install On-Site Testing Facility

Work Remaining:
- Task One: Evaluate Site
- Task Two: Treat Existing Coal Pile
- Task Three: Install Storage Pile Sprinklers
- Task Four: Install On-Site Testing Facility

Plans for next reporting period:
- Task One: Evaluate Site
- Task Two: Treat Existing Coal Pile
- Task Three: Install Storage Pile Sprinklers
- Task Four: Install On-Site Testing Facility

Progress Report 2

Task One: Evaluate Site
- Work accomplished to date
- Work remaining
- Plans for next reporting period

Task Two: Treat Existing Coal Pile
- Work accomplished to date
- Work remaining
- Plans for next reporting period

Task Three: Install Storage Pile Sprinklers
- Work accomplished to date
- Work remaining
- Plans for next reporting period

Task Four Install On-Site Testing Facility:
- Work accomplished to date
- Work remaining
- Plans for next reporting period

Figure 12.3
Several organizational patterns are appropriate for reporting progress. Progress Report 1 emphasizes the overall status of the project; Progress Report 2 emphasizes the status of individual tasks.

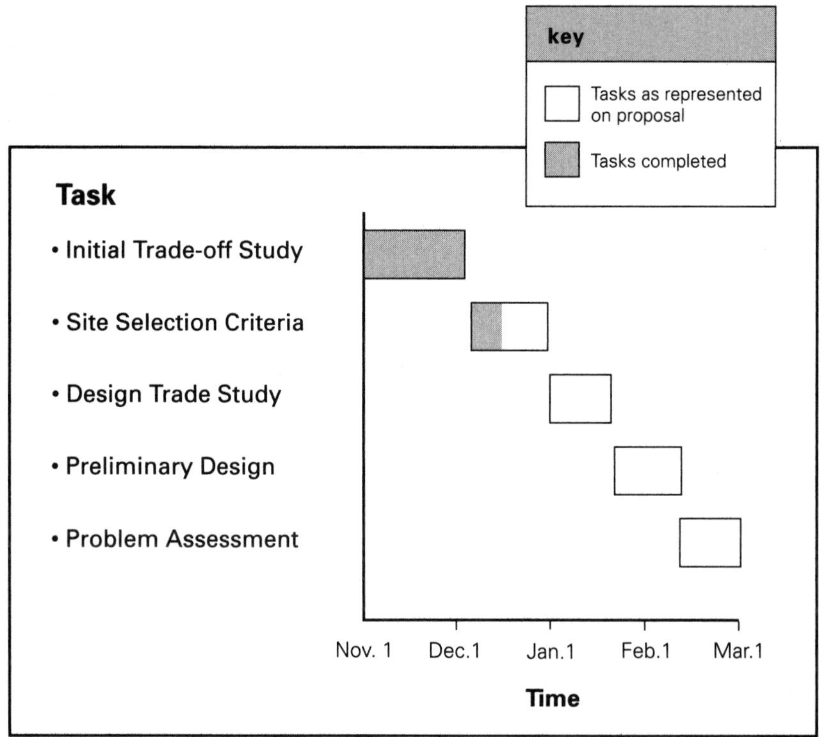

Figure 12.4
Good graphics for progress reports show what was supposed to be accomplished and what actually happened. They serve as planning tools for the research group as well as for project sponsors.

Design and Distribution

Progress reports often have more graphic than narrative information. Figure 12.4 displays the tasks defined in the proposal and provides a rapid view of the status of the project, showing which tasks are completed and which are not. Establish one style for graphics to chart and track your progress. Use the same graphic forms for each progress report in a series.

Many widely available software packages can be used to create spreadsheets, tables, and figures, so that monthly or quarterly reports can be easily compiled by updating the last report. Photographs provide vivid and dramatic information about progress on some kinds of projects. Many organizations now ask for electronically submitted progress

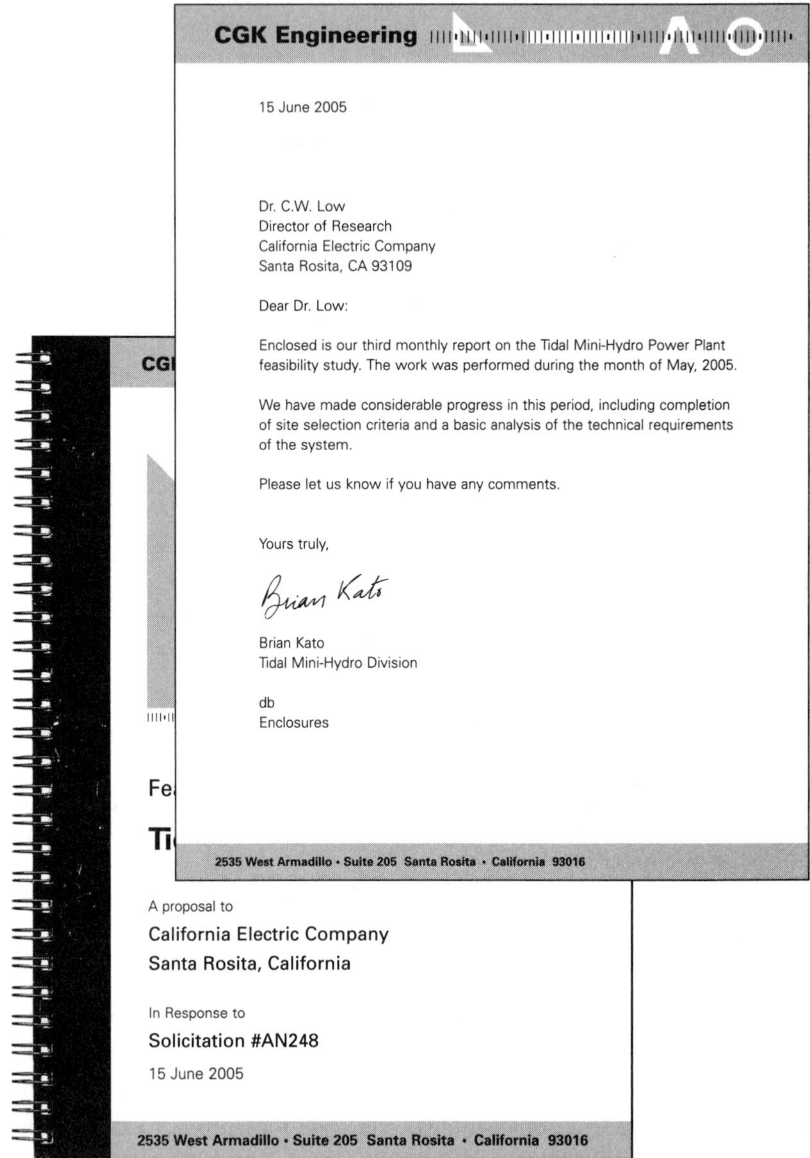

CGK Engineering

15 June 2005

Dr. C.W. Low
Director of Research
California Electric Company
Santa Rosita, CA 93109

Dear Dr. Low:

Enclosed is our third monthly report on the Tidal Mini-Hydro Power Plant feasibility study. The work was performed during the month of May, 2005.

We have made considerable progress in this period, including completion of site selection criteria and a basic analysis of the technical requirements of the system.

Please let us know if you have any comments.

Yours truly,

Brian Kato

Brian Kato
Tidal Mini-Hydro Division

db
Enclosures

2535 West Armadillo • Suite 205 Santa Rosita • California 93016

CG
Fe
Ti
A proposal to
California Electric Company
Santa Rosita, California
In Response to
Solicitation #AN248
15 June 2005
2535 West Armadillo • Suite 205 Santa Rosita • California 93016

Figure 12.5
Progress reports should be carefully prepared, attractively packaged, and, for external sponsors, accompanied by a letter of transmittal.

reports, with video and audio attachments that give the nervous sponsor, several thousand miles from your laboratory or construction site, more information about your progress.

Though progress reports are intermediary documents, they are not throwaways. Create a functional and attractive package for your report. Progress reports with external audiences should have front and back covers. Some companies provide their sponsors with loose-leaf binders in which to store numerous reports in a series. For progress reports submitted outside your own organization, include a letter of transmittal, just as you would with a proposal or a final report. In Figure 12.5, a progress report is accompanied by a brief letter giving major highlights of the reporting period.

Document Databases

Because you write progress reports while you are actively engaged in scientific and engineering work, you will want to streamline the process, reducing the amount of time you need to spend on each report by adopting a functional document format, establishing a style for project-tracking graphics, and creating computer files of standard information, text, and graphics. Some text sections—perhaps a safety report or a weather impact assessment—will be required in nearly the same form in every report in a series. Instead of typing these chunks of standard text each time, you can record and save them as separate entries that can be quickly tailored and inserted into new documents as needed.

A progress report is, by definition, not the last word on anything. But it has a crucial role in the continuum of documents that track the life cycle of a technical project, measuring what happened against what you hoped would happen, providing opportunity for rethinking and negotiation. And a progress report provides content and direction for the final report, which will mark the conclusion of a project. When you have finished a series of progress reports, you may find that you have already written a substantial part of the final report.

13

Reports

■

As manager of a team busily finishing a funded research project on a tight deadline, you know that the final report requires careful planning to be completed on time. You've discussed the report requirements with other team members, and you've called a meeting to establish a schedule for completion. Group members have met to consider the scope of the report, and they know exactly what sections they are responsible for.

Because you have distributed a style guide for text and graphics, sections written by multiple authors will be assembled without significant revisions at the last minute.

In science and engineering literature, the term *report* describes a document that presents results. A report may assess project feasibility, provide observations and commentary on an inspection trip, specify a design solution, or evaluate environmental impact—for just a few examples.

A report may appear within another document, perhaps a memo or a letter, but larger formal reports are prepared in book format, with title page, table of contents, lists of illustrations, sections or chapters, and appendixes. The formal, final report contains some of the familiar elements of proposals and progress reports, but the emphasis is always different. A proposal says, "This is what will happen"; a progress report says, "This is what has been happening and what is expected to happen next." A final report says, "This is what happened."

Reports on the Writing Continuum

Just as a progress report is never the last word on a subject, a final report is rarely the first word. The goals and achievements of a technical project have probably been written up in laboratory notebooks, meeting minutes, proposals, or progress reports.

In the continuum of written work that proposes and records scientific and engineering activity, some elements are unchanged from document to document. A review of related research, for example, might be essentially the same in the proposal and the final report. Some elements reflect only a change in emphasis: A final report may account for the time management of a project, but it is unlikely to focus on this area as much as a proposal must. In a final report, some elements are eliminated: A final report rarely contains curricula vitae of investigators, for example, but a proposal usually does.

Availability of Reports

In most cases, a proposal has very limited circulation, while the availability of progress reports that track the status of the funded work is somewhat larger but still limited. Completion reports, in contrast, may be widely disseminated, though never so widely as journal articles. Technical reports originating in private industry are usually proprietary documents, circulated internally and not available to outsiders. But reports that result from federal grants will be indexed in the National Technical Information Service (NTIS) database; the NTIS electronic catalog contains 400,000 titles (⟨http://www.ntis.gov/⟩). Additional online and printed reference services that provide access to reports include Engineering Index, Chemical Abstracts, and the National Aeronautics and Space Administration's STAR index of scientific and technical aerospace reports (⟨http://www.sti.nasa.gov/Pubs/star/Star.html⟩). In current practice, reports are made accessible to other interested researchers though electronic searches of titles, abstracts, keywords, and even full-document searches. Therefore, the limited circulation accorded a proposal document contrasts significantly with the potentially vast audience for a *nonclassified* final report, which becomes part of the permanent record of what is known on a subject.

Audiences for Reports

Readers of formal reports nearly always represent a complex, varied audience with different purposes, different amounts of time to spend on the document, and different information needs. A cost accountant reading a report that recommends the substitution of geothermal steam for conventional electricity is more interested in the cost sections than in the technical sections that specify details about deep-drilling equipment. An environmental impact analyst may consult only the executive summary, the environmental impact section, and selected appendixes that amplify information about affected wildlife. Very few readers will read every section of a long report. In general, early sections of reports are less technical than later sections, and appendixes are usually directed at specialists.

Reports need to be constructed so that varied audiences, with varied purposes for reading, can chart their own reading paths. Folk wisdom says that 80 percent of readers will read only 20 percent of any document. You cannot therefore shortchange any part of the document. Instead, you need to make each section strong and self-sufficient.

Report-Writing Conventions

Report writing takes a good deal more intellectual activity than following a formula or a recipe. In writing some reports, you will be given a prepared outline—an intellectual template—and required to write standard sections. For other reports, you will need to devise a structure that suits your purpose, the needs of your audience, and your subject. Figure 13.1 lists conventional elements for research reports on scientific and engineering subjects.

Technical reports usually have the same three-part structure we've seen in proposals and progress reports: front matter, body, and appendixes. Within these three broad divisions, reports vary widely in choice of elements and degree of emphasis on any one element. While a student laboratory report may emphasize the way an experiment was performed, a professional research report will focus on analysis and implications for future work. A management report is more likely to emphasize conclusions and recommendations for action than methodology. An environmental impact report will follow a set of guidelines strictly prescribed by the U.S. Environmental Protection Agency, focusing on alternatives to the proposed action as well as environmental consequences.

Front Matter

Letter of Transmittal The letter of transmittal accompanies the report, identifies the item being sent, and provides a context. The letter also provides a brief overview of report contents, typically emphasizing findings of general interest (Figure 13.2).

Cover The cover should indicate names of authors, date on which the report was submitted, organization or institution in which the report was prepared, report number or other indication of the occasion for the report, and proprietary notices if they are appropriate. Many organizations have preprinted forms for use as cover pages, with identifying address and logo.

A report title should name the subject in as straightforward a way as possible. The title should serve as the report in miniature for the widest possible audience. Construct titles with informative words. In some on-

Technical Report Format

Front Matter

- Letter of Transmittal
- Cover
- Abstract or Executive Summary
- List of Figures and Tables
- List of Abbreviations and Symbols

Body of Report

- Introduction
- Theory
- Experimental Section
- Results
- Discussion
- Conclusion
- Recommendations

Appendixes

- References
- Supporting Details

Figure 13.1
Conventional elements in formal research reports.

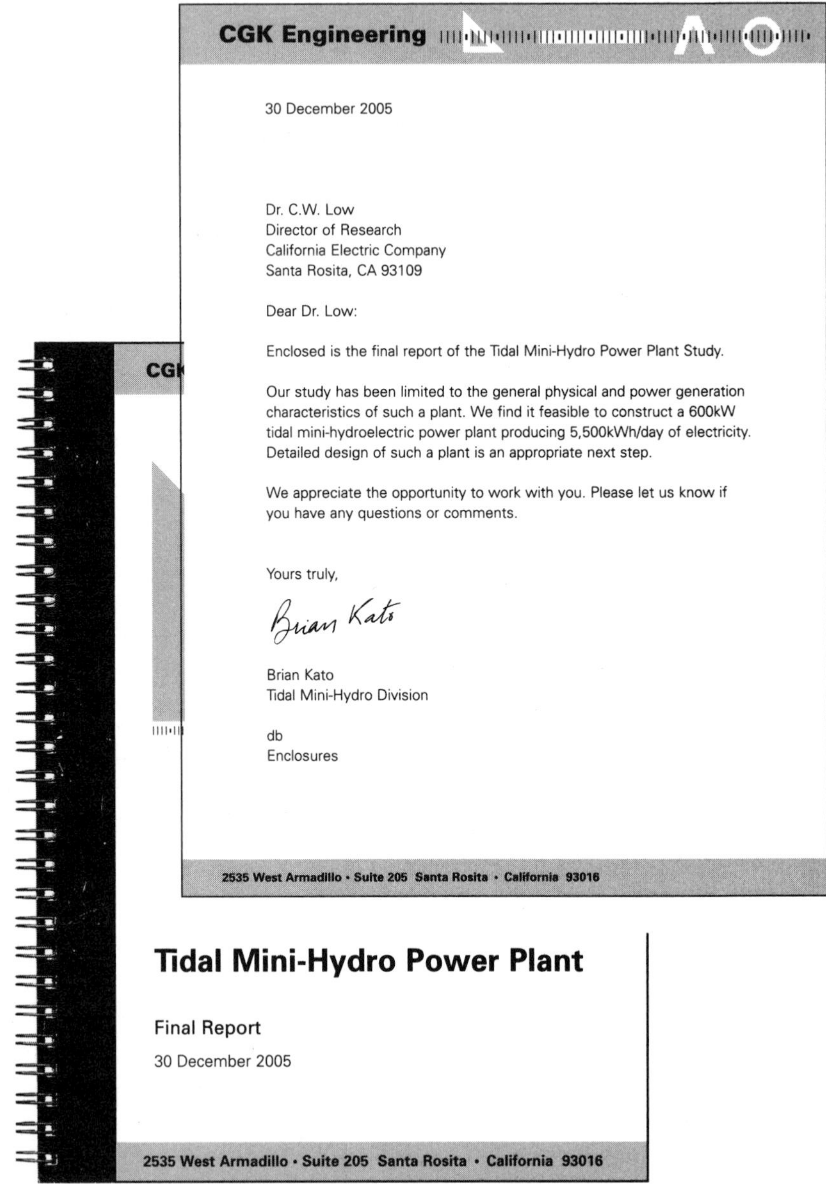
CGK Engineering

30 December 2005

Dr. C.W. Low
Director of Research
California Electric Company
Santa Rosita, CA 93109

Dear Dr. Low:

Enclosed is the final report of the Tidal Mini-Hydro Power Plant Study.

Our study has been limited to the general physical and power generation characteristics of such a plant. We find it feasible to construct a 600kW tidal mini-hydroelectric power plant producing 5,500kWh/day of electricity. Detailed design of such a plant is an appropriate next step.

We appreciate the opportunity to work with you. Please let us know if you have any questions or comments.

Yours truly,

Brian Kato

Brian Kato
Tidal Mini-Hydro Division

db
Enclosures

2535 West Armadillo • Suite 205 Santa Rosita • California 93016

Tidal Mini-Hydro Power Plant

Final Report

30 December 2005

2535 West Armadillo • Suite 205 Santa Rosita • California 93016

Figure 13.2
This letter of transmittal provides a context for the reported work and a summary of contents.

line databases, a keyword list is electronically generated from your title, so you need to build a title from words that represent the most important concepts in your document.

Titles can often be improved if you eliminate inessential detail. The title "Survey and Evaluation of Electrical Power Sources as to Their Potential Application with the Controlled Airdrop Cargo" is overloaded with words that have little information value. A simpler version cuts the number of words and focuses on content: "Potential Electrical Power Sources for Controlled Airdrop Cargo."

Abstract Most scientific and technical reports contain an abstract, a concise account of the problem addressed and the results. The format of abstracts sets them off from the rest of the document: They are typically written as a single paragraph and printed single-spaced, indented on both sides.

Abstracts are classified in two types: informative and descriptive. Informative abstracts are typically about 150 words long, and they present methods, results, conclusions, and recommendations of the report in miniature. An informative abstract frequently stands for the entire report; it may contain all the information that nonspecialist readers want to know about your research (Figure 13.3). Descriptive abstracts are often no more than a sentence, and they may not go much beyond the information already presented in a title. Unless you have been given explicit instructions to the contrary, you should prepare an informative rather than a descriptive abstract for any report.

Abstracts are most profitably drafted after the report has been written. Think of an abstract as a smaller document that describes, but does not evaluate, a larger document. Unlike an executive summary, an abstract does not need to simplify technical concepts or sell a subject. The audience for the abstract is likely to have the same level of technical understanding as the audience for the full report. The tone of an abstract is objective, not persuasive.

Executive Summary Management reports begin with executive summaries rather than abstracts. The executive summary is pitched at readers who may lack the technical expertise to follow particulars of the work but are interested in the implications of the report. In the

Winward, Alma H. 2000. **Monitoring the vegetation resources in riparian areas.** Gen Tech. Rep. RMRSGTR-47. Ogden, UT: U.S. Department of Agriculture, Forest Services, Rocky Mountain Research Station. 49p.

This document provides information on three sampling methods used to inventory and monitor the vegetation resources in riparian areas. The vegetation cross-section methods evaluates the health of vegetation across the valley floor; the greenline method provides a measurement of the streamside vegetation. The woody species regeneration method measures the density and age class structure of any shrub or tree species that may be present in the sampling area. Together these three sampling procedures can provide an evaluation of the health of all the vegetation in a given riparian area.

Keywords: riparian sampling, vegetation cross-section, greenline, woody regeneration

Figure 13.3
An informative abstract presents a condensed version of the report. It is self-contained, with no references to illustrations or appendixes.

foreground is concern with what the results mean and what actions should now be taken.

Executive summaries, often five to ten pages in length, contain the ideas of the report in semitechnical terms, and they replicate the format of the report in miniature, with headings and illustrations. Sometimes they are detached from the body of the report and bound separately for distribution to a much wider audience than the report will have (Figure 13.4).

The tone of an executive summary is typically persuasive, sometimes even enthusiastic, with emphasis on the importance of the problems addressed. Executive summaries stress benefits, conclusions, and recommendations rather than the way the work was done. An executive summary should be self-contained and complete, so that readers are not directed to pages of the report but have all the information they need to understand the main findings.

Table of Contents The table of contents is an important section of any report, even a short one. Include primary and secondary headings, thereby giving readers a quick overview of report contents. A table of

Value Engineering with Existing Infrastructure

An Example Using Wind Energy on the Antioch Bridge

EXECUTIVE SUMMARY

Final Report
December 22, 2006

Mel Manalis, Ph.D.
Jim Davidson
Environmental Studies Program
University of California, Santa Barbara
Caltrans Contract 53Q386

Introduction

The two mile-long Antioch Bridge spans the San Joaquin River as it flows westward towards the Carquinez Straits in a wind-swept environment. The bridge structure amplifies the wind speed as the air flows around the bridge. We were intrigued with the idea of exploiting this amplification of wind speed to produce electricity from wind turbines mounted on the bridge. Would bridge-mounted wind turbines be able to produce electricity from this flowing air at an unusually low cost and yet be compatible with bridge structural and environmental requirements?

This study identifies the technical, economic, and environmental issues germane to answering the above question. We directed considerable effort toward determining the spatial and temporal variations of the wind speed surrounding Antioch Bridge. We began anemometer studies on June, 1988 to search for and document areas suitable for electricity production from wind turbines affixed to the bridge. Subsequently, about 18,000 hours of wind energy data from several anemometers located on and near the bridge have been recorded and analyzed.

1

Discussion

1. Wind turbines are to be located at areas of maximum wind energy underneath the bridge, away from auto, maritime, and foot traffic.

2. The amount of wind-generated electricity that can be produced annually from all the amplification zones under the entire bridge is 26 times the electricity consumed by the toll plaza headquarters.

3. Eighty percent of the electricity is generated from the bridge-wind turbines in the summer, when the utility's demand for electricity is the strongest.

4. Revenues, costs, and investment incentives associated with wind-generated electricity at the bridge indicate potential for considerable cost savings.

5. Environmental impacts of bridge-attached wind turbines involve avian activity but give no cause for concern. No threatened and endangered species occur on the island.

2

Recommendations

1. Interact with Caltran's Office of Structure and Design to explore the bridge-turbine interfacing issues delineated above.

2. Develop new contracting possibilities with representatives of PG&E regarding offsetting electricity at the toll plaza headquarters with electricity generated from wind turbines located under the bridge. Perhaps these discussions could address future electricity demand for lighting the bridge.

3. Examine the compatibility of bridge-supported wind turbines with the Sherman Island Wildlife Management Plan.

3

Figure 13.4
The authors of this executive summary include an overview of their project, a brief discussion of findings, and three recommendations. They prepared the executive summary as a separate bound volume, and they included relevant maps (not shown here). (Courtesy of M. S. Manalis and J. Davidson.)

contents can also help you construct and reconstruct cogent documents. The headings provide a structural view of the document. They may reveal organizational defects that would not be easily obvious in line-by-line reading.

List of Figures and Tables If you have two or more figures or tables, provide a list with full captions. Technical readers often determine whether a report will be of interest to them by scanning the list of figures and tables (sometimes called list of illustrations). Like tables of contents, these lists are also helpful to you as you revise your writing. A review of the captions in your list of illustrations may reveal defects in logic or organization. Sometimes, you can find better ways to present your findings by relocating illustrations or writing more informative captions.

Lists of Abbreviations and Symbols Though a list of abbreviations and symbols (also called a glossary) may not be required, you should consider whether your readers could use such help. In many fields, these lists of terms are essential; even active researchers can hardly keep up with the growing number of abbreviations and acronyms. When you provide a glossary or list of abbreviations, however, you still define each term the first time you use it in your report.

Report Body

Introduction The introduction to a formal report has three functions: (1) to define the problem addressed, frequently with a review of previous work in the area; (2) to state explicitly the objectives of the present work; and (3) to summarize the main conclusions and applications of the work. Introductions may be one paragraph or several pages long. Though problem definition is the key to a successful report, all three functions of the introduction are important. Experienced report readers scan introductory sections for sentences with openings like "In this report, we examine . . ." They know they will find in such sentences a concise statement of topic, objectives, and findings.

Theory Authors of research reports may develop and present their own models, or they may rely on other published work, providing citations to the papers that developed the original models. The solutions for most problems do not require you to develop basic principles but to apply old ones. A theory section is not always needed. However, if the theoretical structure on which your report is based is innovative or particularly involved, the theory should be presented. When you rely on models developed in other published work, cite the papers that developed the original models and establish why you are citing them. Take care to review the literature accurately and carefully. Simply listing reference numbers at the end of a sentence can raise doubts about your use of sources. Be sure to explain what you are applying from each source.

Experimental Section If an experimental section is relevant to your report, use it to describe the laboratory equivalent of your problem: the tools and processes that enabled you to meet your stated objectives. Here you convert your concept of the problem into the language of the laboratory, so that readers may test your methodology against your results in their own laboratories. Clarity and accuracy are priorities. You are describing a variety of objects, materials, processes, and instruments that, used in a specific way, must deliver a characteristic set of data.

In all technical description, an overview helps readers grasp the purpose and scope of your experimental work. Complex procedures are more effectively described when they are arranged in discrete subsections. Consider providing a drawing for any complex apparatus you intend to describe in detail. And for complicated preparation processes or experimental procedures, consider placing the detail in an appendix.

Results The results section translates your findings into the terms of closely observed phenomena, the data of instruments, and the language of numerical generalization and statistical analysis. All the discussions of your report either lead up to or away from this often brief section. Results sections often retain their value long after the methods and conclusions have become obsolete.

Data may be presented in many ways. If the results are simple, note them in a prose passage. Present series of data in tables or graphs. Avoid simply noting that the data are shown in a given table or figure. Rather, draw out an important point or two about the trends shown in each table or graph and emphasize these points again in titles or legends. If your methods of reducing data or estimating their accuracy are based on other published sources, provide the reference.

Discussion The purpose of the discussion is to evaluate results. Results do not speak for themselves; you must interpret them. Interpretation should return the reader to the original objectives stated in the introduction, as well as to the initial theoretical discussion.

Begin by amplifying the most important findings and noting any significant discrepancies. Most authors also discuss the reliability of their main findings. If you can identify errors in your own work, you lend credibility to your discussion by noting them. Even when there is no clear explanation for inconsistencies or errors, you should note their existence. To develop a discussion more fully, you may compare your results with those of other sources.

Conclusion The conclusion section restates main findings, summarizes results in light of the original problem, and draws generalizations supported by those results. The conclusion may contain suggested applications for research results, or it may connect results to other scientific issues. Sometimes the conclusion is combined with the recommendations section.

Recommendations In reports written for management, recommendations sections occupy a prominent place, while in technical reports, recommendations are usually a final, brief section, sometimes combined with the conclusion. You may or may not wish to make recommendations about directions that future research on your topic should take. If the main objective of your report has been to recommend a specific course of action, you will naturally devote an entire section to outlining a concrete and operational set of moves.

Appendixes

The appendix (plural: appendixes) contains information that is so excessively specialized, so lengthy, or so unwieldy that placing it in the body of the text would interfere with reading. Appendixes frequently contain reference sections, maps, glossaries, computer printouts, photographs, extended descriptions of methods, and lengthy comparative data. Appendixes are both self-contained and closely connected to the body of the report.

If the appended information is of more than one kind, create two or more appendixes, each identified by letter and title, for example:

Appendix A. Glossary of Terms
Appendix B. Site Maps
Appendix C. Project Cash Flow
Appendix D. Equipment Specifications

Number the pages within each appendix with the appropriate prefatory letter (page B-3, for example, will be the third site map of Appendix B). Number figures, tables, and equations with the letter designation of the appendix in which they appear (see Figure 13.5).

A good set of appendixes is not a dumping ground for leftover figures and calculations but the location for particularly complex and specialized data that will be crucial for some readers—probably the most important readers. Make explicit connections between appendixes and report body. For example, when you discuss a proposed wind energy farm, tell readers that site maps are located in Appendix A and equipment specifications in Appendix B. List the title of each appendix in the table of contents and consider preparing a second table of contents to material in the appendixes, located just before Appendix A.

Methods in Academic Laboratory Reports

Reports prepared in professional research and development settings usually focus on results, but in academic laboratory settings, researchers and students need to write about how they got the results. A laboratory report usually provides more space for an account of how you did it than of what it means, though a good laboratory report will always draw conclusions and suggest interpretations.

Typical elements in academic laboratory reports are displayed in Figure 13.6 and can serve as guidelines for format. Like any other technical document of more than a page or two, a laboratory report will be improved by the addition of a cover page, a table of contents, a list of illustrations, and one or more appendixes to present important but

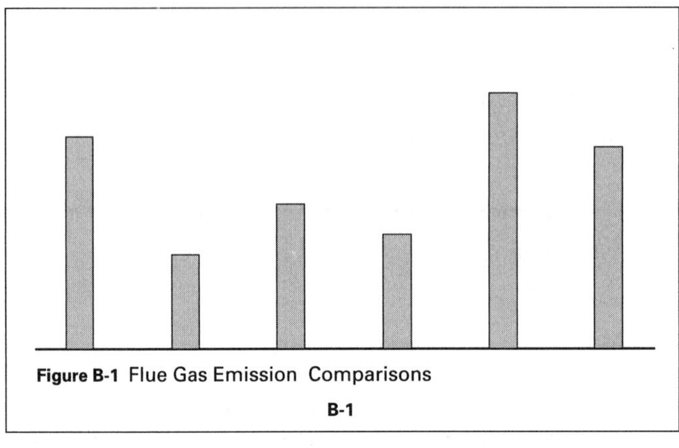

Figure B-1 Flue Gas Emission Comparisons

B-1

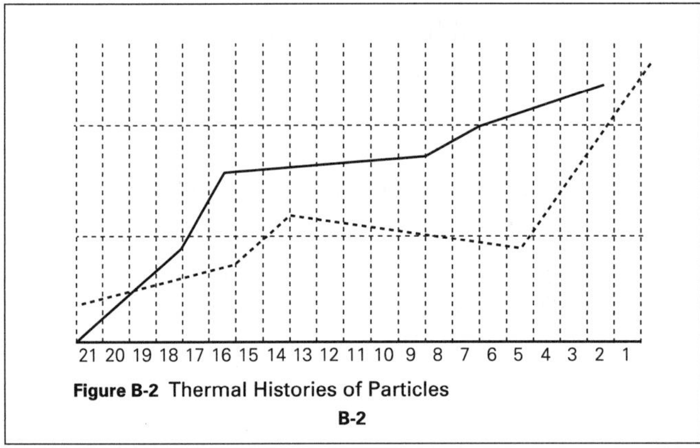

Figure B-2 Thermal Histories of Particles

B-2

Appendix C: Equipment List

- Steel fence posts
- Hammer or post pounder
- Clip board and forms
- Tally counter
- Camera and film
- Plant identification book
- Two 3-foot rods
- One 6-foot pole for use in sampling woody species regeneration
- Flagging
- Global positioning system unit (if available)

C-1

Figure 13.5
Appendixes generally contain information of interest to specialist readers. They are paginated in a different style from the body of the report and referred to within the report.

Laboratory Report Format

Front Matter

- Title Page
- Abstract
- Table of Contents
- List of Figures and Tables
- List of Abbreviations and Symbols

Body of Report

- Introduction
- Theoretical Background
- Equipment List
- Experimental Procedure
- Discussion

Appendixes

- References
- Supporting Details

Figure 13.6
Conventional components of laboratory reports.

unusually detailed or specialized data like computer code or extended calculations.

Decision-Making Processes in Recommendation Reports

Recommendation reports evaluate a process or a product and recommend (or perhaps advise against) a specific course of action. Typically, they begin with an explicit statement of what they recommend. They then specify the criteria that have served as the bases for judgment, and they compare alternatives against criteria. Finally, they list the action required to implement the recommendation (Figure 13.7). Recommendation reports are the stock-in-trade of consulting engineers, who use their special expertise to make informed recommendations to others.

Because recommendation reports serve as the bases for decisions, they must provide explicit information about the criteria that have informed the author's thinking (see Table 13.1). What are they? Why do they matter? In a report recommending the location for a new manufacturing plant, for example, criteria for comparing potential locations might include proximity to railroad tracks, state and local tax rates, and available skilled labor pool. A comparison of competing alternatives against the same criteria results in a tradeoff analysis. Candidate Plant Location A may be conveniently located with respect to railroad tracks and an outstanding pool of available skilled labor. The tax rate for A, however, may be so prohibitive that Candidate Plant Location B, with its low tax rate, is preferable despite its great distance from railroad tracks and a marginal pool of skilled labor.

For many problems, one course of action will not be clearly superior to another. In such cases, do not attempt to oversimplify; a frank discussion of your concerns will be more welcome than an attempt to conceal discrepancies. In some research settings, authors of recommendation reports are required to include a subjective rating of their confidence in the chosen alternative.

Emphasis on Alternatives in Environmental Impact Reports

A formal environmental impact report (EIR) is written in strict compliance with specifications authorized by the U.S. National Environmental

Recommendation Report Format

Front Matter

- Letter of Transmittal
- Cover
- Executive Summary
- Table of Contents
- List of Figures and Tables
- List of Abbreviations and Symbols

Body of Recommendation Report

- Recommendation
- Introduction
- Decision Criteria
- Analysis of Alternatives
- Conclusion
- Action Required

Appendixes

- References
- Supporting Details

Figure 13.7
Conventional components of recommendation reports.

Table 13.1
The authors of this recommendation report considered three design options for a converse-piezoelectric pump and rated each on six specified criteria. The summary table they have prepared is helpful to their sponsor because it includes line drawings as well as a rating scheme of plus and minus for each option. (Courtesy Frank Sager and Craig Speier.)

Criteria	Design 1		Design 2		Design 3	
	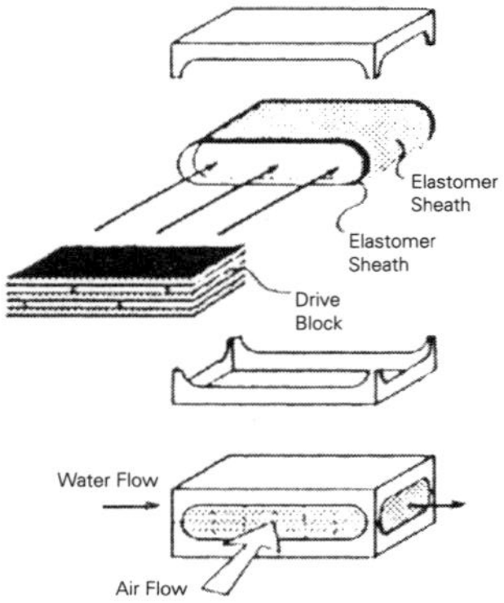 Elastomer Sheath Elastomer Sheath Drive Block Water Flow Air Flow		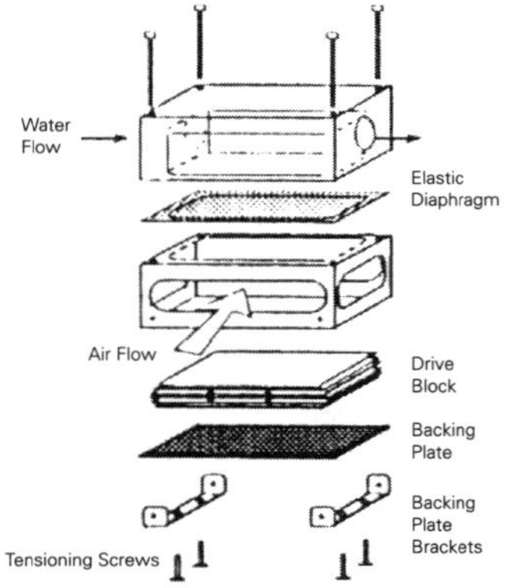 Water Flow Elastic Diaphragm Air Flow Drive Block Backing Plate Backing Plate Brackets Tensioning Screws		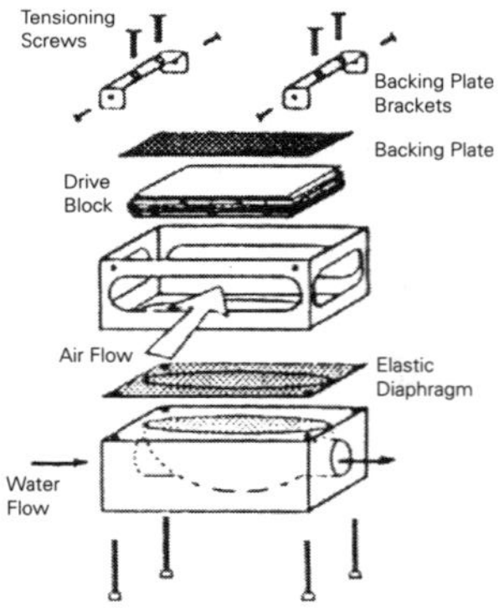 Tensioning Screws Backing Plate Brackets Backing Plate Drive Block Air Flow Elastic Diaphragm Water Flow	
Ease of fabrication	–	Difficult to fabricate the sealing surfaces around the plastic retainer.	+	Easy to make sealing surface with O-ring groove.	+	Easy to make sealing surface with O-ring groove.
Fabrication cost	–	The cost in machining time would be high for this design due to the accuracy required to insure a good seal for the retainer.	+	Less precision required for fabrication of sealing surfaces and less machining time.	+	Less precision required for fabrication of sealing surfaces and less machining time.

Experimental parameter control	–	Diaphragm tension not adjustable.	+	Adjustable bottom plate allows variation in diaphragm tension.	+	Adjustable bottom plate allows variation in diaphragm tension.
Hydrodynamic considerations	+	Good flow characteristics over and under the elastomer sheath.	–	Water flow cavity not hydrodynamically optimized.	+	Water flow cavity hydrodynamically optimized.
Failure mode considerations	–	Any leak in this system will result in water contamination of the drive block and the possibility of electrical short circuit.	–	Any leak in this system will result in water contamination of the drive block and the possibility of electrical short circuit.	+	Since the drive block is located above the fluid cavity, there is less chance of flooding and shorting.
Gravity considerations	–	Hydrostatic force on system at all times is working against imposed force.	–	Hydrostatic force on system at all times is working against imposed force.	+	No hydrostatic force on diaphragm.

Policy Act (NEPA). The distinctive focus in an EIR is on what is lost and what is gained with any decision. A discussion of trade-offs and compromises is, then, at the heart of an EIR (see Figure 13.8). Authors not only need to consider environmental impacts of proposed actions but also to discuss reasonable alternatives to avoid or minimize adverse impacts. A database of final EIRs as well as of draft statements provides helpful models of these important and complex reports (⟨http://tis.eh.doe.gov/nepa/docs/docs/htm⟩).

Managing Complex Report Writing and Production

Planning for Coauthorship and Deadlines

Managing the group writing process for a final report is much like the process of proposal management. Report writing will present the same task allocation and time management problems. The importance of a well-thought-out, section-by-section document plan should be obvious to any research group that has got this far.

There will be a deadline, a time by which the report must be finished and delivered. Schedules, calendars, shared understandings, and frequent group meetings are just as important here as in the completion of any other complex task. The contributions of group members need to be coordinated.

Groups should meet regularly through the research work and should consider the writing required as well as the progress of the technical work at every meeting. When the time comes to write the final report, no group member should have a clean slate, needing to start from scratch to construct sections of the document. Although no report will write itself from assorted laboratory notes and other records kept by project investigators, a report can be more efficiently and effectively written if investigators do not separate the writing tasks from the other investigative tasks of the project.

Distributing Writing and Format Guides

Many active research and development settings have an official style guide available for everyone to use (Figure 13.9). By consulting the style guide, engineers, scientists, writers, editors, technical illustrators, and others involved in document production do not need to create and learn

Environmental Impact Report Format

Front Matter

- Letter of Transmittal
- Cover Sheet
- Summary
- Table of Contents
- List of Figures and Tables

Body of Environmental Impact Report

- Purpose of Action
- Alternatives Including Proposed Action
- Affected Environment
- Environmental Consequences

Appendixes

- List of Preparers
- List of Agencies Receiving Reports
- Supporting Details
- Index

Figure 13.8
Conventional components of environmental impact reports.

Nuclear Division
Document Preparation Guide
Contract W-7405-3ng-26

Chapter 6
Page 6-65

SUBJECT: Preferred Usage

hectare	ha
henry	H
hertz (singular and plural)	Hz
hexagonal close-packed	hcp
high voltage	hv
horsepower	hp
horsepower hour	hp•h
hour	h
hundredweight	cwt
hyperfine structure	hfs

Figure 13.9
Writers in the nuclear division of one government installation refer to the *Document Preparation Guide* distributed by the publications manager. The guide covers a wide range of issues, including abbreviations and punctuation.

a new set of rules for each project. If your report follows from a group of earlier documents, you will want to plan carefully, so that the style guide for the final report has the same features as earlier documents on the subject. Standard publications practices may take time to develop initially, but the long-term payoff is saved time and improved consistency of style and appearance.

The most essential task in document design is to make sections and pages predictable. Even a simple document template will produce the same kind of information, in the same place, in the same type style on every page. (See Chapter 7 for guidelines on page and document design.)

Editing for Clarity and Accessibility

Some sections of reports have a life of their own. As you are writing, remember that titles, keywords, and abstracts may be indexed in technical databases and become part of the literature of science and engineering, independent of the report document. Executive summaries will not be indexed, but they often have a much wider distribution than the document itself. Sometimes they are printed and distributed separately, the only section that influential readers will read. Needless to say, all report sections need attention and care. But sections that will stand alone should be the focus of particularly thoughtful preparation and vigorous editing: They must pass the tests of clarity and self-sufficiency.

Many report writers deliberately invest the largest amount of editing time on the most widely read report features: titles, tables of contents, abstracts, executive summaries, headings, and captions to illustrations. Consider nonverbal as well as verbal features when you edit. Consistent use of white space to create separations between major ideas will increase the readability of your report. You may want to add tabbed dividers to increase the ease with which readers can find exactly the information they need, or you may use color to highlight key points in complex illustrations.

Bringing All Drafts and Boilerplate Up-to-Date

By the time you write a final report, you probably have both text and illustrations that you can update and reuse. If you have labeled and maintained data files carefully, you may save yourself needless duplication of effort. But allow time to review previously written material (often called *boilerplate*) with great care: A mechanical cut-and-paste effort will not be enough to assure currency and consistency. Figures and tables will always need to be renumbered from document to document, and text references to figures and tables will need to be updated accordingly. Heading styles may need alteration for consistency. Verb tenses usually need attentive editing from proposal to final report.

Using Headings to Map Report Structure

Because headings stand out from the text, they send powerful signals about the relative importance of the material that follows. Their size, placement, and typographical features are therefore important. Establish a style for headings and always take the time to edit a report for consistency in heading style.

Even generic section and subsection headings like Background, Process Description, and Conclusion will help nonspecialist readers recognize parts of a report. But reports aimed primarily at nonspecialist readers can be more effective if headings and subheadings are informative. For example, "System Efficiency and Maintenance" is more specific than "Technical Criteria." When the exact wording of headings is presented in the table of contents, readers have an easily accessed, helpfully detailed overview of the report.

Making Graphics Accessible

Tables and figures—collectively called illustrations in scientific and engineering writing—take up a large percentage of the pages in most reports. In fact, many reports contain more pages of tables and figures than of text. Every illustration is a piece of technical literature in its own right, one that may be more widely studied than the text of the report and that may actually circulate separately and be reused in another report.

Every illustration should have a caption, set above a table or below a figure. Captions should be informative—many report readers skim the text but read illustrations and accompanying information with extreme care. Caption style should be consistent throughout a report: either full sentences or sentence fragments.

Every table and figure should be numbered. For long reports, illustrations are usually numbered according to the section of the text in which they occur (Figure 4.2, for example, is the second figure in the fourth section). In shorter reports, figures and tables are usually numbered straight through. All tables and figures that are derived from sources must mention their source at the bottom of the illustration, set close enough so that the illustration never gets reproduced without the reference (see Figure 9.2).

All tables and figures should be referred to with the word "Table" or "Figure" and the designated number. Some writers like to integrate the reference into the report text; others prefer to use parentheses:

The design takes advantage of a commercially available array of solar collectors (see Figure 6.12).

or

The design, illustrated in Figure 6.12, takes advantage of a commercially available array of solar collectors.

Studies of the ways that technical readers read reports show that parentheses seem to serve an important function, guiding readers back to the place they left in the text when they moved ahead to look at an illustration.

Placement of illustrations presents extreme challenges. The conventional advice is that illustrations should be located directly after the first text reference, but this directive is often not very useful. By the time you have placed Table 2 or Figure 3, you usually find that text and illustrations are no longer on the same page. If you are committed to text and illustrations being closely connected, you can use storyboard format, consistently presenting text on the left-hand pages and illustrations on the right, or you can prepare a separate volume of illustrations so that readers can read text and illustrations side by side.

Informing Readers about Errors

If you discover serious errors in your text or illustrations after the report has been printed, prepare an erratum sheet (for one error) or an errata sheet (for more than one). List the page number and any other information that will help locate the error, such as the line number or figure number (Figure 13.10), and slip the sheet into the report document, between the front cover and the title page.

Erratum Sheet

On page 19:

Table 6, Ocean Disposal of Low-Level Radioactive Wastes, 1957–1969

should read:

Table 6, Ocean Disposal of Low-Level Radioactive Wastes, 1967–1969

Figure 13.10
This erratum notice alerts readers to an error in the report text and provides corrected information.

Publications Beyond the Report

A final report is rarely the first document on a subject, nor is it always the last. In the course of work in science and engineering, research results may be repackaged and disseminated in oral and written forms: at meetings and conference presentations; in conference proceedings, reports, and refereed journal articles. Chances are that portions of your report will appear in other documents, and your results will form the basis for further research.

14

Journal Articles

Targeting a Journal for Submission
Conventions of Refereed Articles
Front Matter
Body of Article
End Matter
Manuscript Preparation
Submission and Resubmission
Collaboration on Journal Articles
Electronic Journals

■

An academic research team spends three years on a project and, with some changes in direction, reports results that please both the team members and the funding agency. With the work complete, most of the team plans to move on to other projects. The team leader, however, has one more project task in mind: final results should be published in a reputable scientific journal. Publication will take little effort, the leader assures the rest of the team. All that remains is to rewrite the final report so that it meets the specifications of a journal article. Promotions, the team leader points out, may depend on this venture.

Advancement in science and engineering is frequently tied to publication of research in refereed journals, where articles submitted for publication are reviewed by several experts ("referees") who assess the validity and originality of the work. With journal publication, your contribution is now in the formal domain of technical literature. Once published, your work is accessible through secondary sources and becomes part of the knowledge in a field. After publication, your article may be cited by others working in the field. Colleagues may evaluate its contribution and link new information to it. Your published journal article can become a reference from which new theory is advanced and new evidence added.

The first release of information on any technical subject is unlikely to appear in a refereed article. The delay from the start of a project to journal publication may be three years or more, so that by the time a journal article appears in print, the authors have probably discussed the research in a variety of written and oral forms. In written form, the information may have appeared in one or more proposals, progress reports, completion reports, papers for conference proceedings, and in some fields, a brief form of journal publication called a *letter*. The same information will probably have been discussed informally through an e-mail discussion forum and may have been the subject of presentations at professional meetings. This sequence of activities helps authors to shape their work for journal publication.

Furthermore, once an article manuscript has been completed, it is considered a "preprint" and may be circulated to other researchers working on similar projects. Preprints are now usually electronic (e-prints), circulated by way of e-mail or posted on Web sites, increasing the speed with which they reach interested readers. Several professional societies have created preprint sites; see, for example, the chemistry preprint server (⟨http://chemweb.com/preprint⟩) or the preprint server maintained by the American Mathematical Society (⟨http://www.ams.org/preprints/⟩). Some authors do not distribute preprints until the article has been refereed and accepted for publication; others distribute preprints to obtain feedback and suggestions for modifying the manuscript. If you hope for ultimate publication in a prestigious journal, it may be a mistake to post your preprint. *The New England Journal of Medicine* and the journal *Science* will not consider for publication anything that has been publicly available on line, whether on a Web site or from a preprint server.

Targeting a Journal for Submission

Estimates vary, but there may now be 50,000 to 70,000 refereed science and engineering journals. Still, the nature of your research and the scope of your paper will limit rather dramatically the number of journals to

Statement of Editorial Policy

Neurochemistry International is devoted to the rapid publication of outstanding original articles and timely reviews in neurochemistry. Manuscripts on a broad range of topics will be considered, including molecular and cellular neurochemistry, neuropharmacology, and genetic aspects of CNS function, neuroimmunology, metabolism, as well as the neurochemistry of neurological and psychiatric disorders of the CNS.

Remote Sensing of the Environment serves the diverse remote sensing community with the publication of scientific and technical results on theory, experiments, and systems design in remote sensing technology and applications. In addition to original research papers, surveys and summaries of previously published works are welcomed, as are comprehensive state-of-the-art articles.

Figure 14.1
Most journals provide explicit information about topics and types of articles that they consider suitable for publication.

which you might submit your work. Journal titles indicate the general area of interest, and many journals also include more explicit statements of their editorial policy (Figure 14.1). Your chances of acceptance are enhanced if your subject is closely matched to the research interests of the journal. Do not send manuscripts to a journal without first carefully considering whether the scope of your work fits the journal's publishing profile. Always analyze the journal in which you hope to publish.

Though some journals publish only one kind of manuscript, many publish contributions in more than one category. Types of manuscripts may include:

Articles or Research Papers These reports of original research work are usually assessed by at least two independent referees.

Letters These brief communications are used in many scientific fields to present information that is timely and important. They are usually no more than 2,500 words in length, and in the interests of rapid publication, many journals send letter manuscripts to only one referee.

Notes Journal notes elaborate on previous papers published in the journal, present new experimental data, or develop a new theoretical concept. These manuscripts often receive only one review.

Reviews Critical reports survey recent developments in a field and are usually commissioned by the editor.

Letters to the Editor Space is usually reserved for discussion of papers previously published in the journal and for miscellaneous topical issues.

Conventions of Refereed Articles

Writing a high-quality paper for publication in a refereed journal takes more than following a formula or filling in the blanks. The refereed journal article develops as a stylized assemblage of sections, each devoted to a specific kind of content: theoretical, methodological, empirical, and interpretive (Figure 14.2).

Many researchers claim that the act of writing a journal article is a significant part of the science. During the process of writing a manuscript, authors often discover gaps in experiments. The writing then sends them back to the laboratory for more experiment and analysis. Recognize that in the writing of even the simpler elements of your article you may find that you have further work to do.

Front Matter

Title The title will be more widely read than any other part of the article. Titles allow potential readers to judge the relevance of the document for their own interests. Titles also provide indexers with keywords to use as they prepare subject indexes for bibliographic reference services. Your title should be concise and informative, reflecting the specific content of your work, emphasizing keywords, eliminating filler words or abbreviations. For example, rather than "Survey and Analysis of HIV-Induced Immunodeficiency Caused by Programmed Cell Death of Reactive T

Journal Article Format

Front Matter

- Title
- Abstract
- Keywords

Body of Article

- Introduction
- Review of the Literature
- Theoretical Section
- Experimental Section
- Results
- Discussion
- Conclusion

End Matter

- Acknowledgments
- References
- Appendixes

Figure 14.2
Standard elements in a journal article reporting research.

Cells," your title might read "Programmed Death of T Cells in HIV-1 Infection." Though there is no standard length for the title of a journal article, it is likely that a two- or three-word title is not very informative and that a 15-word or longer title could be shortened for improved focus. A two-part title with main and subtitle may be appropriate in some cases.

Abstract After the title, the abstract is more widely read than any other part of an article. It is likely to be available electronically even if the full text of the article is not. The abstract is a standalone piece of technical literature, not an introduction to the article but a capsule version of it, the article in miniature. The vocabulary of the abstract is likely to serve as the basis for bibliographic searches on your subject. An abstract that clearly and accurately represents the problem you addressed, your methodology, and main results will ensure that future researchers in your area learn of your work.

Write the abstract for your article in accord with instructions (or examples) provided in the journal to which you are submitting the manuscript. Abstracts vary in their length and content, but they are typically 150 to 200 words. Some—called descriptive abstracts—give only a general idea of what the article covers. Others—called informative abstracts—include greater detail (Figure 14.3).

Keywords For other researchers in your field, the keywords that identify your subject and focus will be crucial electronic access routes to your publication. The journal in which you publish may ask you to select keywords from a predetermined list, or you may be free to select your own terms. You will naturally want to select words that represent the most important concepts in your article.

Body of Article

Introduction The main function of the introduction is to identify the objectives and rationale for your paper. The introduction argues for the originality, the good antecedents, and current connections of your work. The introduction specifies the problem addressed, summarizes previous

Vol.21:187-194, 2000 **AQUATIC MICROBIAL ECOLOGY** Published March 31
Aqua Microb Ecol

Effects of the zebra mussel on nitrogen dynamics and the microbial community at the sediment-water interface

Peter J. Lavrentyev[1]
Wayne S. Gardner[2]
Longyuan Yang[3]

[1] University of Michigan, Cooperative Institute for Limnology and Ecosystem Research, 2200 Bonisteel Boulevard, Ann Arbor, Michigan 48109, USA

[2] National Oceanographic Atmospheric Administration, Great Lakes Environmental Research Laboratory, 2205 Commonwealth Boulevard, Ann Arbor, Michigan 48105, USA

[3] Chinese Academy of Sciences, Nanjing Institute of Geography and Limnology, 73 East Beijing Road, Nanjing, Jiangsu 21008, PR China

Abstract

A flow-through experiment was conducted on intact cores of sediment from Saginaw Bay, Lake Huron to examine how trophic interactions between filter-feeding bivalve mussels and microbial populations could affect nitrogen dynamics at the sediment-water interface. The zebra mussels used in the experiment removed a large proportion of protozoa and phytoplankton from the over-lying water, particularly heterotrophic nanoplankton (up to 82%), while bacterial populations showed less change. A 3-fold decrease in the protozoan to bacterial carbon ratio corresponded to a 2.5-fold increase in relative ammonium removal rates as estimated from the dark loss ^{15}N-ammonium. Excretion by the bivalves also increased net ammonium flux to the water, thus elevating the total calculated areal ammonium removal rates to about 6-fold over rates observed in the control treatment. These data suggest that filter-feeding bivalves may significantly affect transformation rates near the sediment-water interface by excreting ammonium and altering the microbial food web structure at the sediment-water interface.

KEY WORDS: Nitrogen • Microbial food web • Sediment-water interface • Bivalve mussels

i

Figure 14.3
Here an informative abstract summarizes objectives, methods, results, and discussion.

research, and indicates what the present work will add to what is known about the subject. The last paragraph of an introduction typically contains several sentences that serve as both preview and summary of the subject and findings of the paper. Such sentences might take the following form:

"Here we show that . . ."
"In this paper we describe further investigations of . . ."

Review of the Literature A review of the published literature on a specific topic is a standard element in the introductory sections of proposals, reports, and journal articles. Sometimes such reviews are prepared as independent documents, and they serve as summaries and critical assessments of the main trends in a field, including the controversies, the accomplishments, and the direction of present work. A literature review is always heavily documented and accompanied by a reference list of works cited. Readers must be able to access all items that you refer to in your literature review, whether hard copy or electronic.

Theoretical Section Sometimes you can provide enough theory in the introduction to support the paper and move directly to the experimental section. But in many papers, more extensive discussion is necessary, and you will have to provide a separate theory section. The theory section may contain a predictive model or series of governing equations, a survey of design parameters, or a discussion of assumptions. These discussions frame the topics and variables that will be the main subjects of your experimental, results, and discussion sections.

Many articles and reports do not develop their own models but rely on other published work. In these cases, the authors cite the papers that developed the original models. Thereafter, they draw on equations from the cited papers and assume that readers interested in a full development of the models will refer to the original works.

Experimental Section The experimental section of an article describes the tools and processes that enabled you to meet the stated objectives of the introduction. This section is sometimes called materials and methods, experimental methods, procedure, or experimental apparatus, depending on the stylistic preferences of the journal. The section will be read for at least two major reasons. First, readers will judge how skillfully you have designed the empirical processes of problem solving. Second, readers may test your methodology against your results in their own laboratories. In experimental sections, clarity and accuracy are priorities. You are describing a variety of objects and processes that have been used to deliver a set of data. Include significant numbers, but move detailed analyses to appendixes.

Results The results section translates the empirical terms of the laboratory into the language of numerical generalization and statistical analysis. All sections of a journal article lead up to or away from the results section, and the results section may retain its value long after the methods and conclusions have become obsolete. Results are confined to their own section not only because they manifest a distinct phase of your research but also because readers often like to work their own interpretations on data, perhaps considering alternative conclusions. When results are mixed with interpretations, the integrity of the data can be compromised.

Data may be presented in several ways. If results are simple, you may be able to note them in a brief prose passage. Series of data should be presented in graphic form. You do not need to be exhaustive: Condense data according to standard analytical procedures into meaningful representations of your work. Avoid merely noting that the data are shown in a table or figure: Draw out the importance of trends shown in each illustration.

Discussion In the discussion section, you evaluate your results and their significance. Just presenting results is not enough. Data rarely speak for themselves. The discussion section may note discrepancies in the findings and explicitly discuss the reliability of the results. By identifying inconsistencies and noting their sources, you lend credibility to your work. Even when you have no clear explanation for inconsistencies, you should note their existence.

Conclusion In the conclusion, you can restate your findings and assess their implications. In this section you may specify possible applications of your findings and, if appropriate, recommend directions that future research on the topic should take. These statements should bring you full circle to the original problems and objective of the work. They identify your main accomplishments and connect your work with larger issues.

End Matter

Acknowledgments Most journal articles end with a brief section of acknowledgments. With as few flourishes as possible, thank people (other than coauthors) who contributed substantially to the work. Also thank funding agencies for their support, stating grant numbers, as well as organizations that have aided by providing space, equipment, or supplies.

References Several reference styles are used in scientific and engineering publication (see Chapter 9). Base your style on the guidelines provided in the journal. If these are not explicitly stated, you can derive stylistic guidelines by examining published papers.

Appendixes Though journal articles are usually brief accounts of research findings, they sometimes do contain appendix sections with material such as lengthy experimental procedures, expanded discussion of results, or instrument and circuit diagrams. In the interest of keeping printed papers shorter, many societies now publish this kind of supporting information in CD-ROM format or the Web, but not in hard copy issues of the journal. Information about how to prepare supporting information for publication will appear in the Instructions for Authors page.

Manuscript Preparation

Your chances of acceptance are enhanced if your manuscript is prepared in the style preferred by the journal. Most editors provide detailed instructions for manuscript preparation, often inside the back cover of under the heading "Instructions to Authors" and also on the society's Web site. Here you will find guidance about manuscript style, equation style, number of copies required, reference style, and instructions for preparation of tables and figures. Determine the preferred length for manuscripts by studying the journal and checking explicit instructions. The Society of Tribologists and Lubrication Engineers, for example, recommends that papers not exceed 5,000 words or equivalents in figures and tables, for a total of six typeset pages. If additional pages up to 12 are used, mandatory page charges of $160.00 per typeset page will be applied (⟨http://www.stle.org/offer_of_paper/offer_form.htm⟩).

In all cases, follow journal guidelines. Most journals ask that you submit copy for illustrations as well as lists of figure captions and table titles on separate pages. Some specify size limits for tables and figures; some announce page charges, with extra page charges for color artwork (Figure 14.4). Some journals request full electronic submission; others ask for hard copy manuscript, but when the paper is accepted, they ask the author to send a computer disk containing the final version.

Many publishers now supply authors with electronic instructions and templates. The American Mathematical Society, for example, provides author packages for 27 of its publications. Each package includes instructions, author handbook, style files, templates, and samples (⟨http://www.ams.org/tex/author-info.html⟩). The American Society of Mechanical Engineers provides authors with a required Assignment of Copyright form on its Web site (⟨http://www.asme.org/pubs/copy.html⟩).

Instruction to Authors

Manuscript. Manuscripts will be accepted with the understanding that their content is unpublished and not being submitted for publication elsewhere. Manuscripts should be submitted to the Editor-in-Chief. Four copies are required. All manuscript material should be typed, double spaced, on one side of $8^{1/2}$ x 11 in. white bond paper. Allow margins of at least 1–inch on all sides of the typed pages.

Title. All titles must be as brief as possible, 6 to 12 words. Authors should also supply a shortened version of the title suitable for the running head, not exceeding 50 character spaces.

Affiliation. Include on the title page full names of authors with academic and/or other professional affiliations, and the complete address, including e-mail, of the author to whom proofs and correspondence should be sent.

Abstract. An abstract of 250 words or less should accompany each full-length paper. Avoid abbreviations, diagrams, and reference to the text.

Keywords. Authors must supply 3-10 key words or phrases that identify the most important subjects covered by the paper.

References. Citations in the text are by author(s) and dates in parentheses. Full citations should be arranged alphabetically and must conform to *The Chicago Manual of Style*, 14th ed.

Illustrations. All figures and tables must be submitted in a camera-ready form. Label each with article title, author's name, and figure or table number by attaching a separate sheet of white paper to the back of each. Each figure or table should be provided with a brief descriptive caption; all captions should be typed on a separate page at the end of the manuscript. Do not include tables within the text; type all tables on separate sheets. If color photographs are submitted, a letter should be included stating that funds are available to pay for color reproduction. The publisher will forward an estimate of cost to the author before beginning color reproduction.

Page proofs. All proofs must be carefully corrected, with all editor's queries answered and returned to the publisher within 48 hours of receipt. Corrections of typographical errors are permitted, but the author will be charged for additional alterations to the text at the proof stage.

Disks. Upon acceptance, authors should submit a computer disk containing the final version of the paper, along with the final manuscript. Specify what word processing software was used, including which release, and type of computer used.

Figure 14.4
Most journals provide manuscript preparation instructions to authors. Instructions for preparing artwork and references are particularly important, and they vary among journals.

Submission and Resubmission

Except for commissioned review articles, editors will rarely invite you to publish your research (though this happens). More likely you will submit your manuscript, accompanied by a one-page letter of transmittal addressed to the journal editor (Figure 14.5). In this letter you should state that you want to have the paper—identified by its title—reviewed for publication. In a second paragraph you should present the main focus of the article. Conclude the letter by thanking the editor for considering your request. If a copyright waiver form is required for publication, it should be enclosed with your letter; many journals provide such a form in each issue. Some journals encourage authors to suggest names of several experts in the field who would (or who would not) be good referees, and you can include such information in the letter of transmittal. Some journals ask that you include a statement confirming that the manuscript has not been published previously and is not also being considered for publication in another journal.

The editor will screen the paper to ensure that its subject is appropriate for the journal. At this stage, your letter of inquiry may elicit a simple rejection letter in response: Your paper has not even been reviewed because its subject matter falls outside the range of the targeted journal. More likely, you will have considered this issue before you shipped off your paper, and your manuscript will be sent out to several expert, anonymous reviewers who will independently assess the adequacy of the science, the significance and originality of the project, and the clarity of presentation. Reviewers then return their comments to the editor (Figure 14.6). Sometimes an author must resubmit a manuscript several times before achieving publication, but critical feedback from referees often improves the manuscript.

The editor will respond in one of three ways:

Rejection In a letter of rejection the editor may include excerpted comments from reviewers or a summary of those comments. The comments may be of help in revising your paper for submission to another journal.

Conditional acceptance On the basis of reviewers' comments, the journal is interested in publishing your article but only if certain changes are made. The editor delineates those aspects of the paper that, in the esti-

CGK Engineering

15 March 2006

A.R Nathanson
Editor, *Tidal Energy*
Dept. of Ocean engineering
University of California
Santa Rosita, CA 93106

Dear Professor Nathanson:

Enclosed are three copies of an original paper, "Design Tradeoffs for Tidal Mini-Hydro Power Plants," submitted for review in *Tidal Energy*.

The study compares three basic methods of operating a tidal hydropower system: tide-cycle, ebb-cycle, and double-cycle. The material has not been published, and it is not under consideration for publication elsewhere. The research reported in the paper is part of an ongoing study of the feasibility of tidal mini-hydro power plants.

Thank you for your attention to this paper. We look forward to receiving your comments.

Yours truly,

Brian Kato

Brian Kato

BK:sw

Enclosure

2535 West Armadillo • Suite 205 Santa Rosita • California 93016

Figure 14.5
This letter of transmittal accompanies three copies of a manuscript submitted for journal publication. The letter briefly summarizes the article and also states explicitly that the authors have read and followed the instructions to contributors.

Studies in Earth Science **Confidential Reviewer's Report**

1. Is the subject of the paper consistent with the scope of the journal?	YES	NO
2. As far as you know, has this material been published before in English?	YES	NO
3. Does the scientific content justify the space it will occupy?	YES	NO
4. Can any parts of the paper be shortened or omitted without loss of scientific content?	YES	NO
5. Are any errors of fact or logic contained?	YES	NO
6. Are all figures necessary?	YES	NO
7. If the paper contains graphs and tables based on the same data, is it necessary to include both?	YES	NO
8. Does the abstract (normally 50-150 words) bring out the main point of the paper?	YES	NO
9. Is the title suitable and adequate?	YES	NO
10. Are the literature references adequate?	YES	NO

11. What is your overall recommendation?

___ Publish as submitted **___ Do not publish**

— Publish with major revision **___ Publish with minor revision**

Comments: please use a separate sheet to expand on the above and to suggest changes that you feel would improve the manuscript, in particular with regard to its length.

Signature: ______________________ **Date:** ______________________

Figure 14.6
Reviewers of journal articles typically return a checklist and a separate sheet of detailed comments to the editor. The journal mentioned here is fictitious, but the form of the confidential reviewer's report is widely used.

mation of the reviewers, need revision. This type of acceptance is the most typical. Few articles are accepted outright.

Acceptance On the basis of reviewers' comments, the journal publishes your paper as submitted—a rare occurrence.

If you receive a letter of conditional acceptance, you may decide to follow the editor's comments and change certain parts of the paper, or you may decide to withdraw the paper from further consideration. In either case, you must respond in a timely fashion with a letter stating your decision. If you decide to change your paper, your next submission will be reviewed again, either by the same referees or by the editor alone. You may need to go through several cycles of submission and review before the editor considers your paper ready for publication.

The former editor of *Science*, Philip H. Abelson, took an optimistic view about the chances of a paper being published. Although *Science* accepts only about 20 percent of manuscripts submitted, Abelson believed that almost all of the rejected papers appeared eventually in other journals. Recognize, of course, a difference in prestige among journals. An article rejected by prestigious research journals but finally published in an unrefereed electronic newsletter will have substantially less influence on the practice of science and the progress of a career than if it had been published in the originally selected research journal.

After the paper has been accepted, it may be copyedited for stylistic consistency and correctness by an editorial assistant on the journal staff, and your next task will be proofreading. Journals that levy page charges for publication may ask you to submit them when you return the proofs. You can also, at this time, order extra offprints.

Collaboration on Journal Articles

Research in science and engineering is rarely a solitary task; most published papers have two or more authors. Coauthors need to ensure that each person listed participates fairly in planning, writing, and revising; developing and monitoring a work schedule; and editing to eliminate differences in style.

Coauthors may work in the same laboratory or university, but they are frequently located at great distances from one another. Successful coauthoring usually requires at least an initial face-to-face meeting to brainstorm, plan, and divide the work, with explicit discussion of ways to coordinate activity and monitor progress.

The coauthors may decide to divide the work by giving each researcher primary responsibility for one or more sections of the paper, designating one person with responsibility for combining and editing. Or they may designate one person to write a draft of the entire paper, while all other members of the group function as editors. Or they may give responsibility for tables and figures to one or more group members and responsibility for text to others. Each of these methods works for some groups and not for others. The critical factor in successful collaboration is shared understanding of the purpose and content of the document so that each writer knows the larger context into which a contribution fits.

When a manuscript-in-progress is posted on a Web site, several authors can view the same display on their respective workstations while they work on the same underlying data structure. As a draft of the manuscript is assembled, reviewing editors can make on-line annotations by creating a node with comments linked directly to the section being discussed.

Electronic Journals

The refereed journal system is in transition. Printed journals do not provide adequate speed, access, or economy. Delays from submission of papers to publication mean that refereed, archival journals cannot help researchers keep up with new developments. To anyone doing serious work in an active scientific or engineering field, information in journal articles is relatively old. The volume of published information is immense, and no researcher can hope to keep up with important developments by merely subscribing to a few journals. Furthermore, the costs of production and distribution have created burdens for publishers and libraries, as well as subscribers. Electronic journals therefore speed dissemination of ideas and change social practices.

Will the printed journal be replaced by electronic publications? Electronic journals certainly have advantages: Articles can be speedily and cheaply disseminated; subscribers can receive only the articles they want to read, rather than entire bound issues that may contain only one or two papers of interest. Articles need not be static repositories of outdated thought: As new information becomes available, the original author—or new readers—might add information, perhaps improving a reference section, amending a theory, or adding links to later work.

At the beginning of 1996, there were about 100 on-line, full-text, peer-reviewed journals in science and technology, including medicine. By the year 2002, that number may approach 10,000. *The Directory of Electronic Journals, Newsletters, and Academic Discussion Lists* (Association of Research Libraries, Washington, DC) provides bibliographic access to electronic documents and grows larger in each new edition. The impediments may primarily be human: Authors publish in traditional journals at least in part for the prestige and recognition that lead to professional advancement. As authors of electronic papers, they might lose professional standing, at least temporarily. Yet electronic journals provide convenience and thoroughness for scientists and engineers in their research. Measures of prestige will inevitably evolve; how frequently the electronic version of an article (an e-paper) is accessed may become a criterion for tenure, status, and advancement.

Although imagining a world without printed journals is difficult, computers have facilitated dramatic changes in the way journal articles are written, refereed, produced, and distributed. Writing now routinely includes e-mail contact between collaborators. Citations are derived from computer searches of appropriate databases. The text is produced—and ultimately submitted—electronically. Referee requests and reports are likely to be sent by e-mail or fax. The major limitation on full electronic submission has been that graphics cannot ordinarily be transmitted with the same verisimilitude as text, but that problem is being solved with special reading devices loaded with the text.

The printed journal will probably be around for some time. And if it does become obsolete, the effect on working scientists may be only minimal. Journals do not contain current views on specific research problems, so reading of preprints will continue to be an essential way to keep up in a field.

15

Oral Presentations

■

Your project is nearly finished, and you're pleased with the design and possible applications your group has considered. Now, as you enter the development phase, you will be assembling a new team. Your next move seems obvious: to present aspects of your work at a professional meeting. The exposure for your research will help attract colleagues who might become part of the project. A formal presentation should also enhance your stature as a research scientist. But first, you need to make your presentation clear and engaging.

The life of a project presents numerous occasions to talk: in relatively informal group problem-solving sessions, in briefings or question-and-answer sessions with clients and sponsors, in formal presentations at professional meetings. On these occasions, you are likely to be in the same room with your listeners, although advanced communication technology has made physical distance between speaker and audience increasingly irrelevant.

Engineers and scientists often approach oral presentations with anxiety. Perhaps they should. Talks can be hard, even impossible to follow. Slides or overhead transparencies can be too complex to read in the few minutes they are displayed. Talks scheduled for 10-minute time slots sometimes go on for 15 or 20 minutes, in blatant disregard of the other scheduled speakers and of the audience. Surely no presenter sets out to be boring, obscure, and insensitive, but unfortunately, many are.

No one method guarantees transfer of knowledge directly from your head to the heads of your listeners. Preparing and presenting technical information that reaches listeners means considering factors that are not always easy to assess. You do know, from your own experience as an ear- and eyewitness to unsuccessful presentations, that getting all the words and numbers in is not enough.

Nor is it enough (or even necessary) to have the slick delivery style of a network news announcer or to have professionally prepared multimedia props. Well-constructed content is the crucial factor in presentation success: You have a strong and interesting idea, and you make the presentation fit the listening and learning styles of your audience.

Most of us can remember times when we could not learn from talks because the speaker was incomprehensible. We can probably also remember speakers—often professors—who violated our every expectation about successful communication and yet were comprehensible and even inspiring. Listeners are interested in ideas and techniques that they can take from the talk and apply to their own work. When they attend a talk at which valuable ideas are put forward, they are remarkably forgiving of less than ideal delivery style.

Listening vs. Reading

As difficult as many written documents may be, they are still potentially easier to learn from than oral presentations. Readers can choose their pace and return to sections that demand further study. Listeners cannot go back over what they have just heard, and they cannot ask the speaker to stop so that they can follow a thought of their own. In preparing oral presentations, you need approaches different from those for written documents, and the differences apply to the visuals as well as to the text.

A talk should not be a "speechified" report or journal article. Your presentation must conform to a different model of information transfer. Accommodate your subject to the way people learn from listening: Stick to a few main points rather than to every line of thought connected with your subject. Remember that the audience for an oral presentation is often less specialized than the audience for a journal article. If the overhead transparency with the outline of your talk reproduces the table of contents of your paper, you may have neglected to prune and shape your topic for the benefit of listeners. In Figure 15.1a, this generic table of contents is suitable for a written report but not for an oral presentation. In Figure 15.1b, the table of contents has been rewritten to serve as the slide in the oral presentation.

Professor Patrick Winston, one of the founders of the Artificial Intelligence (AI) Laboratory at MIT, has often given a presentation called "Some Lecturing Heuristics" to overflow crowds of students and faculty. Winston applies the AI concept of *frames* to the construction of talks from which listeners can learn. A frame is a structure for containing data, a blank to be filled in, like a name and address box on an application form. People are likely to come to your talk with some frames ready to be filled about your content and focus. They will base their expectations on your title and abstract and perhaps your previous work.

These generic frames will be modified by a specific presentation. You can facilitate learning by explicitly providing listeners with frames: spaces where concepts can be entered. Start with the general picture, Winston advises. Discuss your presentation; tell the audience what frames you will be filling in, so they will not have to figure out what you are talking about and how your talk is organized. Be explicit: Say things like "By

Desalination of Salt Water Using Wind Energy

Table of Contents

iii

(a)

Figure 15.1
The table of contents displayed here is fine for a written report, when readers can spend as much time as they need going back over unclear points (a). When the author prepared a table for contents for his oral presentation, he used informative rather than generic headings and provided audiences with crucial previews of his three main points (b).

Desalination of Salt Water
Using Wind Energy

☑ Fresh drinking water is an endangered resource.

- U.S. consumption 360 billion gallons/day
- Severe depletion of aquifers

☑ Wind energy can be used to desalinate salt water.

- Important sources of salt water
- Desalination can utilize off-peak electricity.

☑ Technology can be applied at numerous U.S. and international sites.

(b)

Figure 15.1 (continued)

the time we're finished here today, I want to convince you ..." Your talk can then recall frames, fill them in, link them together, and in some cases, change them, or even create new ones.

Your Audience and Environment

To develop an effective presentation, you need to combine what you know about how people learn from listening with what you know about your particular audience. Who are they? How many of them will you face? Why have they come to hear your talk? What do they already know about your subject?

You also need to consider the physical and social environment in which you will speak. What kind of room? What time of day? What has the audience been doing before your talk? What will participants do afterward? Are you the only speaker, or are you competing for time and attention as member of a panel?

The answers to these questions will not always be what you might wish, but you will not be able to keep secrets from your audience. The room may be dark, airless, and poorly soundproofed. The time for your talk may be Sunday morning at 8:30 A.M., on the last day of a four-day meeting. Audiences respond positively to speakers who acknowledge mutual concerns and often with hostility to speakers who do not. If you carry on as though nothing is wrong while your talk is nearly drowned out by the noise from rooms on either side, audience members may well wonder if you care about what is happening to them. In a case like this, you might indicate that you, like them, find this setting unacceptable and that you hope to have informal conversations on your subject with members of the audience at another time, in a more suitable place.

Structuring Your Talk

In presentations, your aim should be to uncover key points rather than to cover every detail. Your talk is probably not the first word on any subject, and it does not need to be the last. A written version may be available to your audience in the conference proceedings or as a handout. Organize your talk for listeners, not readers. Ask yourself, Given what I know about my audience, what is the clearest and most convincing sequence in which to order the information in my talk?

When you have developed a strong organization, be sure to make it obvious and explicit. The well-known preacher's wisdom is good advice: "Tell them what you're going to say. Say it. Tell them what you said." In successful presentations, you need to develop the technical content of your talk, but you need also to develop the verbal and visual structures that allow listeners to learn from your talk.

Begin technical presentations with explicit discussion of the way your talk is organized. Preview main points. Define specialized terms or key phrases. Tell listeners when you are finishing one section of your talk and starting another ("I've talked about cost factors, and now I want to turn my attention to environmental concerns"). End technical presentations with a review of main points, saving your best formulations for last so that listeners will not be irritated at a mechanical reiteration of what they have just heard.

Selecting a Visual Medium

Because listeners learn and follow better when they have something to look at in addition to something to listen to, visual props are a standard feature of technical talks. In deciding which visual medium is most appropriate, you will need to apply what you know about the audience and environment for your presentation. What visuals do these listeners expect? What visuals will other presenters be using? What visuals will work well in the room? Answer these questions before you determine what is technically possible. A multiprojector, multimedia presentation is not necessarily well suited to all subjects, audiences, or settings. Consider, too, the power of paper handouts, still the highest-resolution means of information transfer. With presentation software packages, it is possible to prepare hard-copy handouts for audiences, with miniaturized or full-size copies of each visual. Be aware, however, that some societies do not permit speakers to distribute handouts during conference presentations.

Chalkboards

Despite its limitations, the chalkboard or markerboard remains a favorite medium, particularly for in-house or academic presenters. It is widely available, and it does not require training or electricity. Because copy is not fixed, the speaker can operate on it: adding, deleting, circling, underlining, color highlighting. Though writing on a board slows a presentation, the delay can also give listeners time to absorb what you are saying. And you can, in some cases, write or draw on the board before your presentation.

On the downside, board space is limited. Rearranging items is time-consuming when they must be erased and redrawn, and items disappear when the space is needed for another topic. Even chalkboards are evolving these days toward electronic forms that combine the features of the traditional chalkboard with the storage and communication links of workstations.

Overhead Transparencies

Overhead transparencies are, for good reason, widely used to support technical presentations. Overheads can be viewed in ordinary room light. They are inexpensive and easy to prepare with or even without presentation software. They are easy to carry and store, and they can be efficiently retrieved during a question-and-answer session.

Overheads are a dynamic medium: Speakers can write on them to emphasize points, cover part of the transparency to progressively disclose the image, and prepare overlay transparencies to be placed over the first. At some presentations, attendees are given a photocopied set of overhead transparencies.

Electronic Slide Presentations

Compared with overheads, slides offer more interesting design possibilities and have a higher visual impact, with outstanding color contrast and accuracy. Many presenters now use standard office software programs like Microsoft's PowerPoint to create computer-based slide shows, with images projected by way of a light valve onto one large screen. Presenters can easily incorporate data from existing files—including spreadsheets, scanned photographs, and Web pages—to create a new presentation. Computer-based slide shows are often posted on organizational or conference Web sites, making it possible for interested audiences to review the presentation and also for others who did not see and hear the original presentation to learn from it anyway (Figure 15.2).

Electronic presentations provide opportunities for dynamic and interactive meetings. Images are often not only displayed but also manipulated, with input from audience as well as presenter. During the course of a presentation, a speaker can move into a spreadsheet program to do some recalculations or access a Web site for display. One matter that conference presenters need to consider is the availability as well as the cost of computer and projection rentals and Internet connections. At some professional meetings, speakers are charged for all such equipment requests.

Producing Effective Visuals

Keep Visual Material Simple

Develop one idea per visual—a single point, relationship, or conclusion—with plenty of blank space. If you need to show details, prepare separate visuals on the same subject, progressively disclosing complexity. Whatever visual medium you choose to support your presentation, assume that you will need to create new copy. Illustrations transferred directly from a technical report to a presentation slide or overhead transparency are nearly always dense and hard to follow.

Design the Right Number of Visuals

A workable formula is that the number of overheads or slides should equal two-thirds the number of planned presentation minutes. Of course, the optimal number of visuals depends on subject and audience, but if you have fewer than one-third the number of visuals to the number of minutes, you are probably trying to put too much information on some of your visuals. Conversely, if you move in the other direction, nearer one visual per minute, some of the visuals are probably not staying on the screen long enough—an effect that can be cumulatively irritating to your audience.

Integrate Visuals

Plan the content and progression of visuals as you plan the organization of your talk—not as a separate process. Many technical professionals use the moviemaker's technique of storyboarding to plan and prepare presentations. Text and visuals are then usually successfully integrated.

Software Agents

Pattie Maes
Software Agents Group
MIT Media Lab

pattie@media.mit.edu
http://www.media.mit.edu

Pattie Maes – MIT Media Laboratory

Agenda

- What is a Software Agent?
- Types of Software Agents
- Roles for Software Agents
- The Future of Agents
- Open Questions

Pattie Maes – MIT Media Laboratory 2

Figure 15.2
MIT Media Lab Professor Pattie Maes prepared these slides for a conference presentation and posted them on her Web site (⟨http://pattie.www.media.mit.edu/people/pattie/CHI97/⟩). Note that the second slide, Agenda, is a preview of key concepts in the talk. (Modified and used with permission.)

What is a Software Agent?

- An over-used term.
- "Agent" theoretical concept from A.I.
- A computational system which:
  - is long-lived;
  - has goals, sensors and effectors;
  - decides autonomously which actions to take in the current situation to maximize progress towards its (time-varying) goals.

Pattie Maes – MIT Media Laboratory 3

What is a Software Agent?

"Software Agent" — particular type of agent, inhabiting computers & networks, assisting users with computer-based tasks.

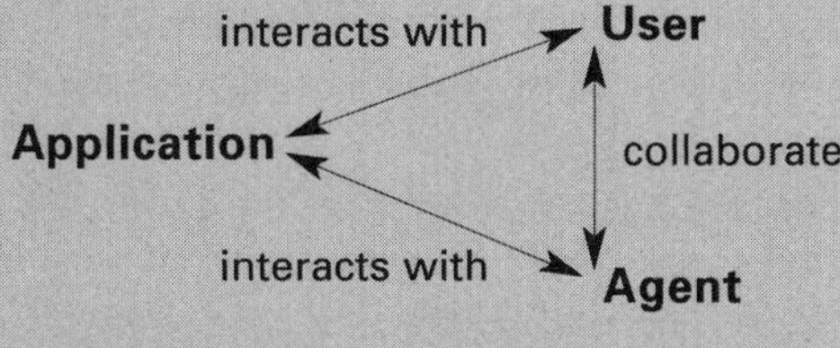

Pattie Maes – MIT Media Laboratory 4

Figure 15.2 (continued)

Types of Software Agents

- Autonomous Robots
- Synthetic Characters
- Expert Assistants
- "Knowbots" or "Softbots"

Pattie Maes – MIT Media Laboratory 5

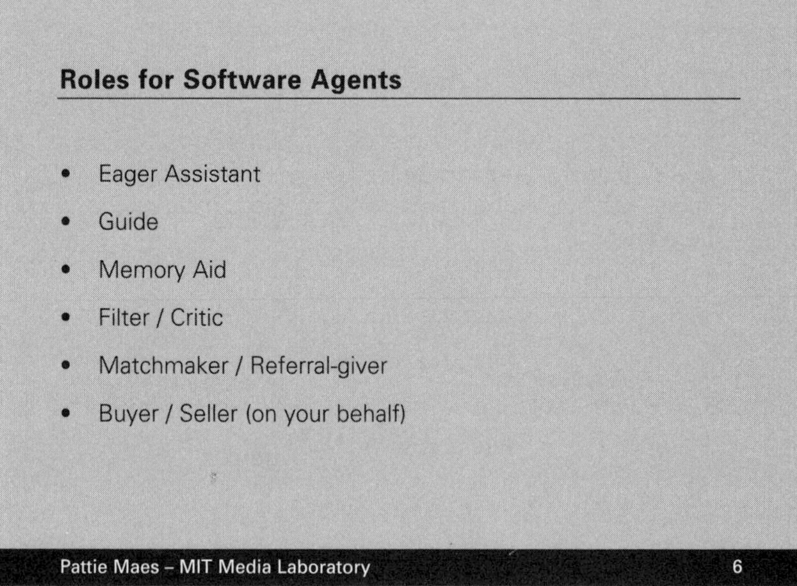

Figure 15.2 (continued)

Future Direction: Markets for Agents
Necessary evolution if we want to keep any bandwidth to ourselves!
agents paying other agents for services
deceitful agents; honest agents
specialized agents built by different vendors
agents advertising services for / to other agents
Pattie Maes – MIT Media Laboratory
7

Open Questions
Privacy issues are key!
Personification: good or bad?
Should agent augment bad habits or enforce "better" ones upon user?
Avoid running amok
Effects on society
Pattie Maes – MIT Media Laboratory
8

Figure 15.2 (continued)

Software is now available to support the storyboarding process, but a do-it-yourself version works well. For a 15-minute talk, prepare an outline and then work with at least 12 pieces of $8\frac{1}{2} \times 11$-inch paper in horizontal orientation. Each sheet represents both the text and the visuals for each minute or so of the talk. Plan the first two sheets as word charts: The first should display the title of the talk and your name; the second should display the overall outline of your presentation. For the remainder, divide each page in two. Use the left side to jot down rough notes for the narration and the right side to sketch the content of the accompanying visual. Tack or tape the completed storyboards to a wall so that you can preview the entire presentation as a unit. With this global preview, you can search for duplications and omissions. You can insert word charts to mark each transition in your talk. You can create a memorable concluding visual. This technique provides better feedback than separate assessment of text and visuals. It more closely models the two streams of information that your audience will be receiving.

Rehearsing Your Talk

When you practice your talk, try putting yourself in the audience's place. Consider how the audience can learn from what you are saying and what you are showing. Consider how much time audience members will need to read and learn from each visual. Don't forward a slide so quickly that it can't be read. Don't block your own visuals as you advance them. Use a pointer tool, which is less distracting than the shadow of a presenter's finger. Remove visuals you are no longer talking about, and turn off equipment you are no longer using.

Prepare and practice every element of your presentation. The standard instructional tool in workshops to improve presentation style is videotape, enabling you to see and hear yourself as others do. But you can learn quite a lot without a video preview. Time yourself: If the talk is too long, cut before you present it, not as you are giving it.

Rehearse with visuals: Practice board work, manipulating transparency overlays, and using computer-based tools to write on slides. Look critically at your overheads or slides from a position in the back of the room. No audience has ever been glad to hear a speaker say, "I know you can't see these slides, but what they show is . . ." If you feel you have no option but to display a visual that can't be seen, provide the audience with a photocopy to accompany your discussion.

Should you ever read your presentation? Practice will vary, and at academic conferences, some speakers read their papers. The preferred presentation style, however, is well prepared but conversational, prompted and accompanied by overheads or slides, perhaps cued by note cards, not full-size manuscript sheets. If you do need to read your paper, start out by not reading: Talk directly to the audience for the first few minutes, setting a context for your paper, adjusting your opening remarks to fit the situation.

Prepare to handle feedback. In question-and-answer sessions, repeat the question so that everyone can hear it. Audiences are frustrated when listening to answers without having heard the question. Do not over-respond; you can offer to continue specialized conversations at another time.

Conference Presentations

Three specialized forms of technical communication are associated with talks given at professional meetings: presentation abstracts, poster sessions, and papers written for conference proceedings.

Presentation Abstracts

Conference organizers frequently ask potential participants to submit a presentation abstract, essentially a proposal to give an oral presentation or a poster session. Presentation abstracts are usually written many months before the meeting, and they may describe work you have not completed. Still, they need to be informative, detailed, and as complete as you can make them. Many professional societies provide Web-based electronic forms on which to submit presentation abstracts, and some societies (the American Geophysical Union, for example) require a submittal fee. Abstracts of accepted papers are frequently published in hardcopy form and also posted on a conference Web site, widely available to all meeting registrants. The audience for your abstract may be large, including those who do not attend the talk and those who do.

Poster Sessions

In poster sessions, speakers are given a bulletin board on which to display graphics and text for a specified period, perhaps two hours (Figure 15.3). Information about the size of the poster board will be available in the conference call; a 4 × 8-foot display board size is common. At many conferences, session managers are available in the poster room to assist presenters with mounting their posters. These sessions are often lively and productive discussions between the presenter and a small, interested audience.

Proceedings Papers

Some conference organizers publish proceedings containing copies of papers given at the meeting. The proceedings may be distributed at the meeting, with text of papers that have not yet been delivered, or they may be published at a later time, with the text presumably revised in light of feedback and discussion. They may be published in hard copy *and* CD-ROM or CD-ROM only. Some proceedings are produced and distributed by the publishing industry, others by conference organizers. Except when they are published after the meeting by an established technical publisher, papers in proceedings are not usually refereed or edited.

Papers in proceedings tend to be shorter than papers in journal articles, not fully developed, and sketchily documented. Journal articles, subjected to rigorous peer review, including textual editing, are more polished. Many speakers who have written papers for conference proceedings later rewrite their findings for submission to a refereed journal.

Science and engineering librarians call conference literature gray literature (*graue Literatur*), because it is often distributed in an unconventional way and therefore difficult to locate. This is not a denigration of the form: In some new fields, much of what is known is available only in conference literature. Conference proceedings are covered in *Engineering Index*, *Science Citation Index*, *Biological Abstracts*, and *Chemical Abstracts*, as well as in specialized publications like the *Index to Scientific and Technical Proceedings* (Institute for Scientific Information).

CGK Engineering

Tidal Mini-Hydro Power Plant

15 February 2006

Abstract

Tidal Hydroelectric Power Plants utilize the difference in head created by the cycling of tides to power a turbine. These low-head turbines convert the mechanical energy of the water pushing against the turbine blades into electrical energy in a generator. Tidal power generators have many advantages. Tidal power's main advantage is its low cost, due primarily to low maintenance requirements and an absence of fuel costs.

Introduction

Tidal power uses minimal land because the majority of the land required is already subject to tidal flood. Tidal power is a renewable resource that produces energy with low environmental impact. Our limited fossil fuel resources are becoming increasingly scarce, and we must search for new and better sources of energy if we are to maintain or improve our current standard of living in this country.

Methods

A reduced-size, fully operational test model of a Tidal Hydroelectric Power Plant was constructed by our design and engineering team to evaluate the structural, cost, and design efficiencies of this system if it were to be implemented on a full-scale basis. Computer-generated modeling, along with a scale replica and numerous subassemblies, was used to calculate the actual data reported here. This system was compared with various other alternative energy sources, including wind energy, photovoltaic cells, ocean thermal-energy conversion, biomass systems, and concentrated solar applications. Oil- and natural gas-fired power plants were used as a benchmark for comparison on an economic efficiency level. Consideration must be given to the current economic climate and the rise in cost of these fossil resources, improving the overall cost structure of Tidal Hydroelectric generation systems.

Results

Our model generated 2.56×10^{12} watts/meter^2 over a period of 9 months after it was put into continuous operation. While we were unable to attain the economic efficiencies of scale that would result from the construction of a full-size system, our costs were surprisingly minimal. The system produced a large percentage of its power when the moon was full, because of tidal cycles. These factors should be considered when implementing the system on a large-scale basis.

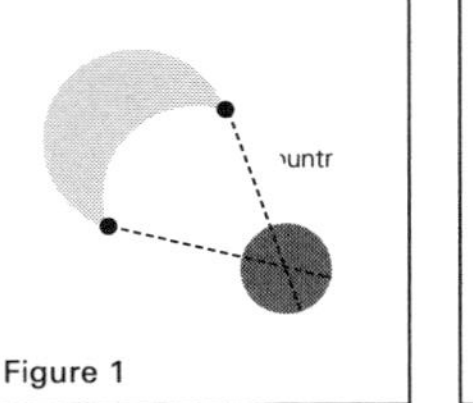

Figure 1

Table 1

Date	Action
January	Design model
March	Construction
May	System testing
June	Data collection
August	On-line model
September	Results

Table 2

Variations	Results
Hydroelectric	123
Solar Cell	456
Nuclear	789
Wind Energy	012
Biomass	345
Passive Solar	678

Results

When compared with alternative energy sources, including wind energy, photovoltaic cells, ocean thermal-energy conversion, biomass systems, and concentrated solar applications, as well as oil- and natural gas-fired power plants, tidal-driven hydroelectric generation systems prove to be highly efficient, as well as economically viable. Very low maintenance costs combined with almost no environmental impact means that these systems could be fully implemented by the middle of this decade.

2535 West Armadillo • Suite 205 Santa Rosita • California 93016

Figure 15.3
This poster is based on a physically enlarged, condensed text of the paper as it might be published in a journal. It includes abbreviated versions of standard parts of the written report.

Telepresence for Meetings of the Future

Improvements in communication technology will make it increasingly feasible to have real-time meetings without the simultaneous presence of participants. In these multimedia settings, new kinds of computer-mediated interactions will be possible, with participants sharing an audio and a visual space. The technical infrastructure of the new conference room will support different models of information transfer—and perhaps better learning. Audiences may be offered less passive roles, and new technologies will change the way people share their ideas.

16

Instructions, Procedures, and Computer Documentation

■

An R&D company has just hired a team to develop manufacturing specifications for a new product. As work begins, the company attorney asks to review the laboratory's safety procedures. To her surprise, the manager discovers that no policies have been drafted. The attorney recommends that work stop immediately. An accident under these conditions would expose the company to serious liability. New employees will sit idle until adequate procedures are written and approved.

Science and engineering present numerous occasions for defining operations—in lengthy documents that exist in their own right and also brief, specialized documents that are parts of longer works. Some, like the methods section in a journal article, will be skimmed but rarely read. Some will be read carefully so that readers can perform the described process. Reference guides may have relatively long lives in service; installation guides may be consulted once.

Instructions are written so that a reader can accomplish something. Procedures, on the other hand, explain how something has already been accomplished. Thus an instruction might tell readers how to work a system, while a procedure would explain how the system works (or should work, in the case of specifications or quality assurance testing handbooks).

Computer technology has enabled significant improvements in the quality, accuracy, and availability of instructions and procedures. Many organizations now deliver such information electronically, storing materials on a central server and distributing them through an intranet. The presentation and delivery medium for instructions can shift from a desktop computer screen to a wireless handheld unit, accessible where needed.

Including Liability and Product Warnings

Procedural writing has some force in law. Poorly written procedures can cause problems ranging from frustration and costly delays to injury and death. An injured worker who had attempted to follow inaccurate or even ambiguous instructions might be able to collect damages for injury. Several liability cases have affirmed the principle that operator's manuals must enable workers to operate equipment safely.

You must clearly and forcefully warn users of all risks and hazards, both with normal use and with possible misuse of the procedure. Correct verbal content is not enough. Warnings and cautions must be placed well in advance of the point they are needed, and they must look different from the rest of the text. Many U.S. military specifications contain good models for safety warnings (Figure 16.1). The American National Standards Institute (ANSI) has developed verbal and visual guidelines for warnings, including signs, safety symbols, and accident prevention tags. The ANSI catalog is available on the World Wide Web (⟨http://www.ansi.org/⟩).

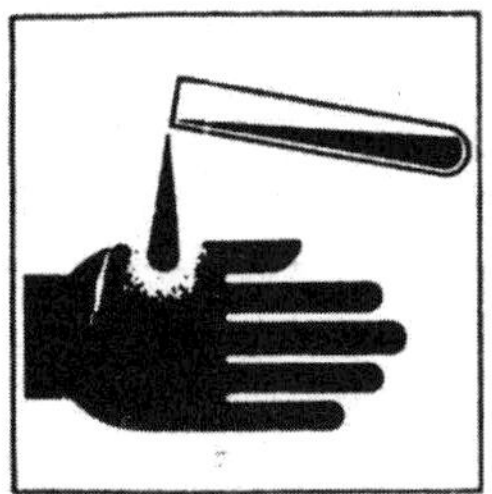
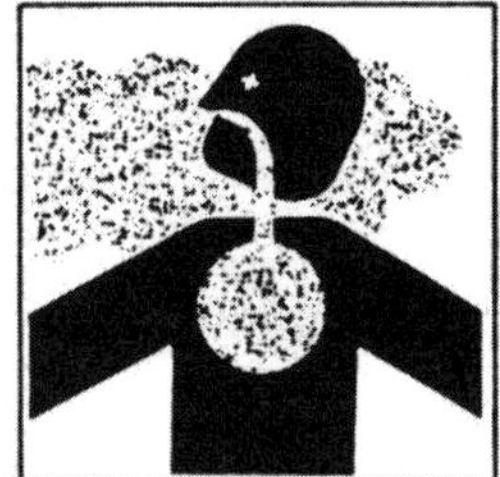

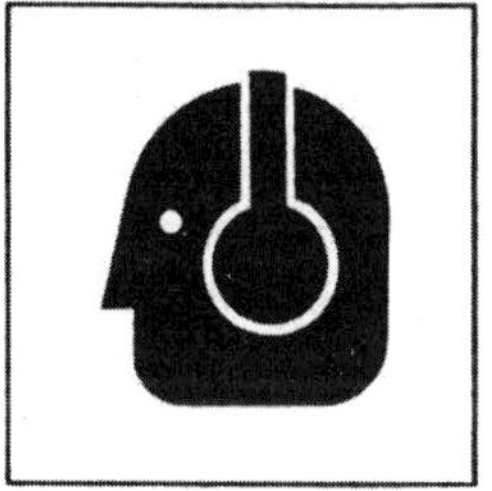

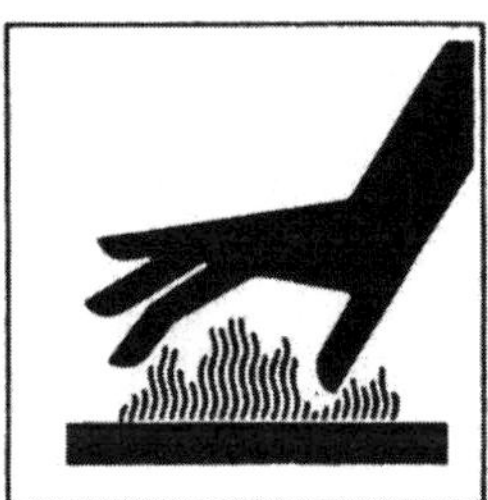

Figure 16.1
These standard warning icons stand out from text and are both dramatic and readable.

Considering Your Audience

Research studies in technical communication, educational psychology, human factors engineering, and information science yield at least one uniform result: People use instructions to get their work done, not to read instructions. They want to find the information they need, and they want to understand the information they find, while spending as little time as possible searching and reading.

When people are learning to do something new, like use a spreadsheet program, they prefer instructions that give them less to read and more to do. They prefer tutorials that give them a chance to practice and accomplish real tasks. Many learn better from documents that have less to read and more to look at. If they come to the task with some prior knowledge, they may be able to complete their work without reading the text, learning only from photographs or line drawings.

Audience analysis is always a central task in document planning. In most cases, you discover that you must address multiple audiences with varied reasons for using your document. Some will need help getting started; others will want to use the product at advanced levels, learning shortcuts and more productive methods for accomplishing their goals. In writing instructions, you need also to analyze the process itself by sorting it into steps. Such a task analysis—sometimes best accomplished by working the process out in rough flowchart fashion—provides feedback on how potential audiences might behave.

Organizing Your Document

When you have pictured the users of your document and their motives and goals, you are better able to organize information to be most helpful to your audience. Remember that while your problem in writing is to decide how you will place and store information, the reader's problem will be to retrieve what you have put there. In what order should you introduce the steps of a process so that readers can follow and learn? Unfortunately, no single answer will work for all readers. Two strategies, however, will at least bridge the gap between your sense of how someone might best learn and the learner's own needs and goals.

First, we recommend that you make your organization visible and explicit. You know what it is, but your reader does not. Reduce the learning burden by explaining how your document works and how you expect readers to learn from it.

Second, provide an alternative path for users who want to create their own information trails. Rather than punishing readers who do not want to follow you step by step through a process, make it possible for them to learn on their own. Some of your readers will choose linear access to your document, following your instructions as you have anticipated. Others will choose random access. Using the table of contents or index, they may jump directly to areas of concern, reading only headings, looking only at illustrations (Figure 16.2). A successful document will enable readers to find what they need in the time they are willing to spend. As a writer, you will need to be flexible, creating a document that accommodates more than one style of reading and learning.

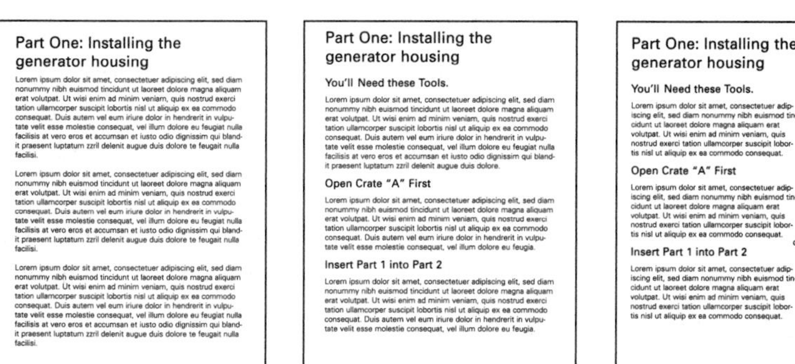

Figure 16.2
Of these three pages of instructions, the first requires reading solid text. The second adds headings, providing the reader with an alternate path through the document. The third adds illustrations with informative captions, giving readers the most freedom to learn and accomplish in their own way.

If you are selecting an organization based on your analysis of the audience and the constraints presented by the procedure itself, here are some familiar strategies:

Alphabetical order
Chronological order
Cause and effect
Order of importance
Spatial order
Division by task
Division by component part

Alphabetical order is a successful organizational strategy for many documents but a poor choice for others. For learning a word-processing system, for example, information organized alphabetically would be relatively useless: a novice could not learn from a list of entries beginning with ASCII files, bold type, caps lock key, directories, endnotes, and so forth from A to Z. On the other hand, alphabetical order is extremely useful for reference guides. Users who know a system will prefer speedy alphabetical access to any topic.

Chronological order is a good choice when the steps of a procedure must be followed in sequence. You might also arrange information by cause and effect or order of importance (simple to complex, increasing to decreasing, most used to least used). Spatial order (left to right, top to bottom) works well when accompanied by illustrations. Division by task and division by component part are patterns that can match function and save readers from skimming an entire document. All of these patterns support a document in which information is stored by the writer in one way. With the addition of an index and informational elements like headers and tabs to indicate what material is on a page, a document with information stored in one way can be used in multiple ways.

Achieving Clarity

You can achieve clarity in instructions and procedures from a variety of strategies, some verbal, some visual, some organizational:

• Give readers advance information about what they will be reading. Informative overviews and headings have dramatic impact on reading comprehension. Readers learn more when they know what they will be learning. They don't have to spend information-processing time trying to determine the topic.

• Divide the operation into modules or segments that allow users to work without turning pages at inconvenient times. Make stopping and restarting easy. For hard-copy documents, if you begin each new module on a right-facing page or on a new spread of two pages, the physical structure of the document will mirror the modular steps of the procedure.

• Establish a consistent way of naming elements in your procedure and stick to it. Decide whether you will say video display terminal, cathode ray tube, or monitor. Do not vary your choice throughout the document. Consider including a glossary of terms.

• Write with verbs that explicitly name the action you want your reader to perform. Figure 16.3 excerpts a very small part of several hundred pages of preferred verbs provided to U.S. defense contractors. This "milspeak" may seem mechanical if you want to be known for a distinctive prose style, but it does enhance clarity. Write in the active voice: instead of "the wheel is to be greased," write "grease the wheel."

• Consider alternatives to conventional linear text. Include, for example, numbered or bulleted lists or message matrixes.

• Minimize cross-references. Readers can follow instructions most efficiently when all of the information they need is provided in one place.

MIL-M-81927(AS)

PREFERRED VERBS

VERBS	DEFINITION	EXAMPLE
Actuate	To put into mechanical motion or action, to move to action.	Actuate the handpump until pressure gage indicates 50 psi.
Adapt	To make fit a new situation or use, often by modifying.	Use the bushing to adapt the fuse to the projectile.
Add	To put more in.	Add electrolyte to battery.
Advance	To make forward; to move ahead.	Advance the throttle.
Wire	To provide with wire, to use wire on.	Wire the circuit.
Withdraw	To take back, away, or out.	Withdraw the bar magnet from the center of the coil.
Wrap	To wind, coil or twine as to encircle or cover something.	Wrap the wire around the terminal.
Zero	To bring to a desired level or null position.	Zero the protractor to the surface.

U.S. GOVERNMENT PRINTING OFFICE: 1975-603-131/1427

Figure 16.3
This list of preferred verbs provided to U.S. defense contractors reduces the creative freedom some writers enjoy, but it also reduces the readers' learning burden.

• Select an appropriate document format. Off-sized pages can be more motivating than the standard $8\frac{1}{2} \times 11$-inch format; spiral bindings permit learners to keep a manual open. On-line documentation needs to be designed *for the screen*, with attention to features that help readers to learn from electronic text.

• Illustrate liberally. Remember that many readers learn better from pictures than from text.

• Accommodate random flipping through pages, the search method that most studies show is still many readers' favorite. Headings, highlighting, and illustrations give a user the freedom to search for what's needed in idiosyncratic ways.

Testing Readability

Technical communicators could use a simple and accurate way to measure the readability of any document. Dozens of readability formulas—some manual, some electronic—have been devised. Most focus on sentence length and complexity of vocabulary as key factors that can be manipulated to improve reading speed and accuracy. The Fog Index (Figure 16.4) is probably the best-known formula.

Skeptics will point out that readability is an extremely complex issue. We certainly agree: Documents acquire readability from a combination of verbal, organizational, and graphic factors, not simply by achieving a

Computing the Fog Index

The Fog Index uses two factors in measuring readability:

1. Average number of words in a sentence (AWS)
2. Percentage of words three syllables or longer (%DW)

0.4 x (AWS + %DW) = Grade level at which text can be read

Figure 16.4

In computing the Fog Index, add the average number of words per sentence to the percentage of long words. Multiply the result by 0.4 for an estimate of the grade level at which the text can be read. (Adapted from R. Gunning and R. A. Kallan. Used with permission.)

numerical score according to a formula. Nor do readability formulas take motivation into account. People will work very hard to interpret extremely difficult prose if they need to and have no alternatives.

So even though the results of a readability test are hardly the last word on the clarity of your document, they are still worth considering. Readability validations are a required stage in complying with many military specifications, and readability software is often supplied with word-processing systems.

Verifying Usability

If your document explains how to accomplish something, go through a full rehearsal of the process as you've written it. Give a draft version to prototypical users for walk-through, testing, and feedback. There is no better way to gather information about the usefulness of an information product or to find defects when they can be corrected.

Document usability is commonly measured through pre- and post-tests, interviews, observation, questionnaires, and read-aloud protocols in which users read a document aloud and express thoughts about it as they attempt to learn from it. Some usability measures are relatively easy to administer and score; others are both complex to administer and time-consuming to appraise. Though none of these methods has absolute validity, each produces feedback about document function, not just about grammar and style.

The ISO 9001 for Procedures

The ISO 9001 initiative of the International Standards Organization has had a major worldwide impact on procedure writing. The initial goal of ISO 9001 was to validate consistency and quality so that products could cross borders within the European Community. To be ISO 9001–certified, a company must document each procedure connected with the production of goods or services. By 1995, nearly one hundred countries, including Japan, had recognized the standards (see ⟨http://www.iso.ch⟩).

Though ISO 9001 standards do not specify document formats, many companies (often with the help of consultants) design templates to be used by procedure writers. Several software packages are available to help multiple authors achieve consistency and clarity. Consistent documentation of procedures is increasingly important for global communication and product development.

Computer Documentation

Most working adults have spent numerous hours learning to use computer systems, and many have formed a poor impression of both paper and on-line computer documentation. In the early days, usability was often constructed through the documentation because it was not available in the product. In recent years, user satisfaction has come to distinguish one computer product from another. Products are less difficult to learn, and the documentation no longer bears the full burden of making technology available to users.

Nicholas Negroponte, founding director of the MIT Media Lab, has argued that the notion of a computer instruction manual is obsolete. The computer itself, he believes, is the best instructor: It knows what you are doing and what you have just done. In his view, software and hardware of the future will come with no printed instructions whatsoever, and the warranty will be sent electronically by the product itself once it feels it has been satisfactorily installed. In the meantime, computer users still require (and expect) excellent documentation: accurate, searchable, and available when needed.

Multiple Information Products

Most hardware and software products require a package of supporting documentation, prepared in several media. Such a package might include an installation guide, a first-day tutorial, a task-oriented guide to advanced features, an alphabetically organized reference guide, a template that fits on the keyboard, and embedded on-line help. These information products can be book based or on-line; they will obviously be more current and accurate if they are available in electronic formats.

Attention to Learning Styles

Effective documentation accommodates users with varied skill levels and learning styles. In well-designed documentation, users can retrieve beginning levels of information without also retrieving advanced levels—

and advanced users can move directly to what they need. Computer users learn better when they can make use of tables of contents, indexes, site maps, headings, previews, and summaries. Whether they are learning from books or on-line documentation, they prefer generous use of white space and markers that allow them to switch attention without losing their place. They do not want to turn pages of manuals or scroll through an online tutorial to locate crucial illustrations.

Task Orientation

Effective documentation is task oriented. Its organization mirrors what users are doing. Novice users need a document structure closely allied to

Topic	Novice Users	Intermediate Users	Expert Users
Getting an Account	✸		
Logging In	✸		
Receiving E-Mail	✸		
Sending E-Mail	✸		
Working with "mbox"		✸	
NNPREP		✸	
Using NN		✸	
Posting with VI			✸
Fingering			✸
Talking			✸

Figure 16.5
Task Matrix for an on-line manual. Many documentation groups begin their work by considering the skill levels of potential users and the tasks each group will want to accomplish. They record their assessments in a task matrix, using it as the basis for organizing, design, and writing.

work they want to do: Headings like "Writing and Editing Your Report" or "Storing and Retrieving Your Document" signal a useful structure. Task analysis is central to preparing helpful information products, and successful documentation is typically the result of extensive task inventories (Figure 16.5).

Accuracy

Effective documentation is accurate. It reflects the realities that the user encounters. As the software is modified, the documentation must be modified, and users must receive updates. For some software and hardware products, developers provide documentation on CD-ROM. Information can be conveniently updated and shipped on a new disk, and the entire volume can be searched electronically. Other developers provide their users with access to a Web or intranet site, including the opportunity to pose technical support questions via e-mail.

The Future of Instructions and Procedures

Advances in computer technology have made it possible to replace printed instructional material with electronic text. The electronic technical manual can be updated as often as necessary, and it is easy to search. Only one copy exists, and users access it on the Web or intranet, where and when they need it: perhaps at a desktop computer or through a wireless information appliance. Information can be delivered in a variety of media, incorporating video, audio, and virtual reality applications. With handheld or wearable computers, field technicians can carry the equivalent of 6,000-page manuals, accessing exactly the information required, reading on a computer screen displayed on one lens of eyeglasses.

In moving to multimedia, the technical skills required to create instructional material have increased. But the benefits are substantial as readers receive the information they need, in forms they prefer: viewing a video to see a step being performed, working a simulation to learn an operation, selecting explanatory links to improve understanding.

17

Electronic Documents

■

Your company has been debating about whether to develop a Web site, and you've been asked to study the issue and write a recommendation. It seems to you that it will be great to have a presence on the Web. You can distribute accurate and updated product information and perhaps even publish white papers describing new products. You might be able to eliminate hard-copy documentation and provide your users with customized multimedia instructions through an intranet. What would it take to produce and maintain a Web site? How can your site accommodate the needs of international users? Can you comply with guidelines for making your site accessible to people with disabilities?

Rather than being bound and fixed on printed pages, electronic "books" are compound documents composed of text, graphics, video images, and audio. Their sequence and even their style of presentation can be selected by the reader. An electronic document can build itself by extracting information from a database and send itself to designated recipients. In settings where security is an issue, electronic documents can conceal contents for users with limited clearance.

On-line communication solves a variety of problems associated with paper. Electronic documents can be customized: Personal training manuals can be created for each learner, based on the trainee's performance on tests of skill. These documents are easy to update, so they can always be accurate. They are easier to search than books, providing improved access to topics and cross-references. They can be remarkably compact: A laptop with CD-ROM drive can deliver a 10,000-page documentation set that would have occupied a dozen three-ring binders.

Links vs. Fixed Paths

Traditional communication is linear. Information is laid out in a single path, and readers move from topic to topic in an order determined by the writer. Electronic information is composed of individual chunks of content and computer-supported links among these chunks (Figure 17.1). Readers follow topics in any order they choose, sometimes guided by a map of the network, sometimes creating their own paths. Elements in an electronic document are in a perpetual state of reorganization. The user can start anywhere and, by way of electronic links, establish connections between multiple kinds of information: text, audio, and video.

A well-designed multimedia system organizes data in a complex, nonlinear way and facilitates exploration of large bodies of knowledge. Each unit can be electronically linked to any other unit, and the user can choose which moves to make. At every step, the user of a multimedia system can see an example or a simulation, look up a definition, listen to sounds, or return to a previous link. Rather than a following a fixed order of presentation, each user, depending on needs and interests, can take a distinctive route through complex material.

Designing Electronic Documents

On-line information needs to be structured for the screen. Displayed pages from printed books will rarely yield effective on-line material. In writing for the computer screen, you must provide for the unique ways users interact with on-line material, facilitating multiple types and levels

Linear Sequence

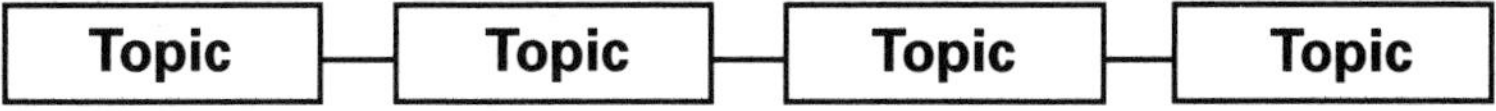

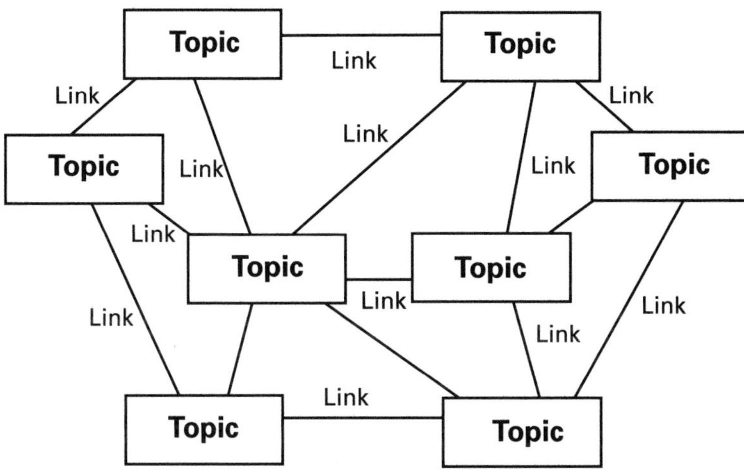

Figure 17.1
In a linear sequence, the order in which readers are expected to learn about a topic is established by the writer. The writer generally assumes that the reader will read the first topic before going on to the second. In electronic text, readers create their own information trails, beginning with a topic and freely pursuing links.

of searches. In a good Web document, readers know what is available and can move efficiently from one topic to another (Figure 17.2).

Many of the principles that guide the production of printed documents apply to the development of Web pages. But Web documents are layers deep—not pages long—and writers must create linked connections as well as chunks of content, all the while exploiting the advantages of graphics, audio, and video.

Provide Navigation Aids

People are used to the physical features of books. A certain heft suggests the time it will take to read the text. Page numbers are visible signals of

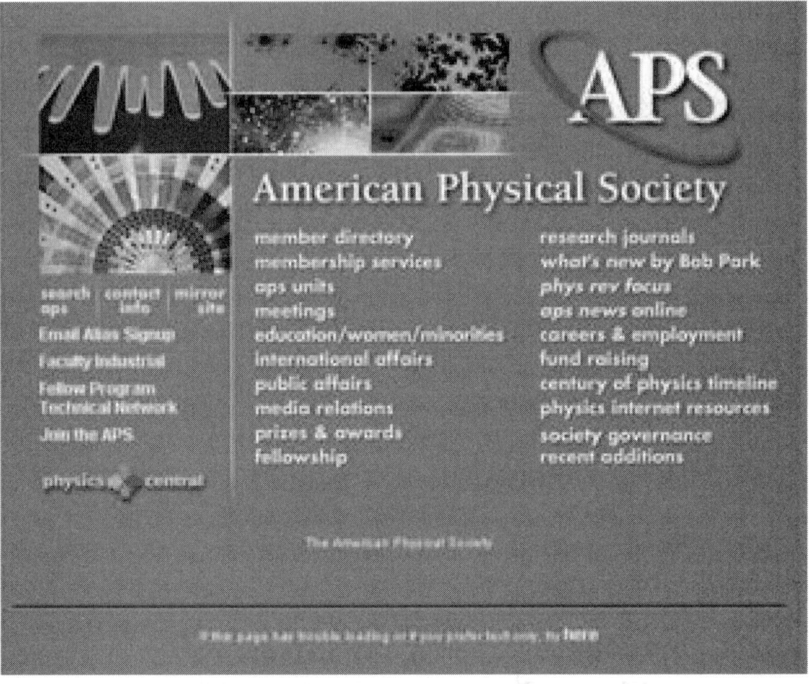

Figure 17.2
This home page for the American Physical Society orients users by providing a table of contents. It also establishes design elements that are repeated through the site (⟨http://www.aps.org⟩).

progress. Pages can be marked and dog-eared. Bookmarks can be placed and replaced. Pages are present even when they are not being read.

With on-line text, users have different cognitive challenges. For most people, moving through several computer screens is not as easy as looking back and forth between pages of a book. On-line information can be confusing. You must structure material to minimize a user's disorientation, providing ways for readers to tell where they are at all times. Organize information in a way that makes sense to users, and provide navigation and escape information on every screen.

Write for the Electronic Page

Conventions for writing electronic text are evolving, but two relatively uncontroversial techniques for improving on-line text involve conciseness and clarity. Write concisely, presenting only small chunks of text to read on each screen. Conventional wisdom holds that readers can deal with seven (plus or minus two) pieces of information at one time. On a computer screen, it appears that a standard of five (plus or minus one) works better to minimize confusion. Substitute bulleted lists for paragraphs, and use white space generously. Use clear and simple language so that readers get it right the first time. Most users of on-line documents do not want to relocate and reread anything. Provide a printable version of your content; many readers still prefer to learn from hard copy.

Design for Consistency and Quick Loading

Designers of electronic documents need to develop style specifications, just as they do for hard copy. A screen has less space than a standard page, and displayed text is almost always less legible than it is when printed. Opinions conflict about which fonts are most legible on screen, which graphic-highlighting devices attract a reader's attention, and what effects color, blinking, sound, and animation will have on reading comprehension. Consistency of design and optimizing of graphics for quick-loading time are crucial.

Create a visual signature for the site, and design all screens in the same format and typographical style. Use a limited number of fonts, styles, and colors. Select fonts that are particularly legible on-screen, and present extended text in 12-point size. Except for headings, use upper- and lowercase letters. Include images that load quickly. Readers of on-line documents expect visuals, but many get impatient waiting for graphics to load, and they move on to other sites.

Copyright Issues

Though some of what is on the Internet is in the public domain and can be copied at will, a large amount of the information on the Internet is protected by copyright. It is best to assume that a work is covered under copyright protection until you have determined otherwise. U.S. courts have decided that developers of a Web site can include on their site links to other Web sites. But many questions of intellectual property in an electronic environment remain unresolved.

Authors of Internet documents can protect their own work by including a copyright notice in the following format: © 2003 Garrett Liu. Authors can register their work with the Copyright Office of the U.S. Library of Congress for a $20 filing fee. Registration forms are available from the Copyright Office Web site: 〈http://www.loc.gov/copyright〉.

Global Audiences

Because the audience for openly-available sites on the World Wide Web is international and multilingual, Web writers do well to think of the ways that their material will be received by speakers of other languages. Many large Web sites are available in multiple language formats, and translation software is widely used. If you write clearly and simply, you can increase the likelihood that your text will be translated accurately. However, Web writers need to think beyond translation to the larger problem of *localization*. When you localize an information product, you adapt it to fit the complex cultural realities of another country.

Visual elements do not predictably transfer across cultures. Users whose written languages are read right to left won't be helped with directional arrows placed at the bottom right of the page and arrows pointing right for the next page. Icons that are widely recognized in one country may appear unclear or even offensive in another. Presentation of dates, times, and orders of magnitude can vary greatly, and you can avoid misunderstandings by mentioning the system of measurement you are using.

Accommodating Disability

Audiences for Web sites are large and diverse and include people with physical limitations such as vision or hearing problems. Most U.S. federal agencies have been required to redesign their Web sites to comply with guidelines that will make the pages more accessible to people with disabilities. Specialized software allows visually impaired users to hear text-based messages and explanations of images. Transcripts or written descriptions of audio clips assist users with hearing problems. The Web site for the accessibility initiative mounted by the World Wide Web Consortium is a good source for announcements of technical advances aimed at providing universal access to the Web: ⟨http://www.w3.org/WAI⟩.

Past and Future Applications

Multimedia has been with us at least since 1978, when the Architecture Machine Group at MIT developed the Aspen Movie Map. This was a surrogate travel application that allowed the user to take a simulated drive through the city of Aspen. A set of videodisks contained photographs of all the streets in the city and some of the buildings. Users could stop in front of many buildings and go inside! The Aspen Movie Map even had a time-of-year knob, giving the user a choice of the autumn or the winter version.

The Architecture Machine Group also created a prototype Movie Manual, suitable for both novice and expert auto mechanics. Among the features of the Movie Manual was this one: Each time a tool was mentioned, the mechanic could link to a picture of the tool and a narrative about how it is used or to a video of an experienced mechanic using that tool.

By the end of the twentieth century, we had already witnessed a dramatic transition from paper to on-line documents. Why provide each of 600 employees with a 500-page manual that needs updates at least twice a year? Accurate and updated information can be delivered on replaceable CD-ROMs, on a proprietary intranet, or on a handheld information appliance connected to the wireless Web.

Enthusiasts are convinced that multimedia applications will be the basis of a new literacy. Software will diagnose a user's abilities and learning needs, and the multimedia book will reconfigure to best suit each reader. Learning will be effective and powerful because, in this view, nonlinear systems model the associative style of human idea processing. Information will always be timely, because electronic updates are cheap and convenient.

Skeptics wonder about a future in which all texts are unstable and can be read in any order, perhaps revised by many readers. Which, if any, versions of a document will be authoritative? What factors of electronic text will substitute for the social signals that distinguish a high-quality printed book from a carelessly prepared handout? What do people need to learn so that they can browse profitably in immense multimedia databases? What will be the long-term effects of nonlinear, multimedia reading? What is the meaning of intellectual property in easily reproduced electronic documents? What is the longevity of digital information? Will electronic documents become obsolete when hardware and software change?

Adult readers are usually more familiar with paper-based than on-line formats, and they do not always know how to learn from electronic documents. Electronic text is less legible, slower, and more tiring to read. Multimedia documents are more time consuming and expensive to produce. But the potential advantages of electronic documents include vast storage capacity, easy search and retrieval, and accuracy.

18

CVs, Résumés, and Job Correspondence

■

Late on a Friday, looking forward to the weekend, you check your e-mail and find an alarming message. You discover that your plans will have to change. The new department manager wants to interview all mid-level staff, and she wants an up-to-date curriculum vitae (CV) by Monday morning. Unfortunately for you, your CV is seven years out of date. You aren't afraid of losing your job, but to be ready for Monday, you will need not only to revise your copy, listing what you've done for the past seven years, but also to update your page design. You've discovered what scientists and engineers know and often ignore: that a current and attractive CV is a crucial document for professional advancement.

Résumé or Curriculum Vitae?

As an ongoing task, you should write and update your professional biography, presenting data in either a curriculum vitae (CV) or a résumé form. Though the terms are sometimes used interchangeably, a CV is a record of academic and professional achievements, while a résumé also includes an employment objective.

Except when you are actively engaged in job searching, a CV will be far more useful than a résumé. Your CV contains the kind of information that conference chairs want as they introduce you to an audience and that funding agencies want when you apply for support. Your employer also may want to see your CV as you undergo a personnel review.

You need to develop and maintain a hard copy version of these documents *and also* one that is optimized for electronic submission and tracking. Computer files are essential for keeping your biography current. Create a file of your most recent CV and a record of each version you have prepared. You can also keep an ongoing file of material to incorporate into the next version you prepare, together with names and addresses of current and potential references.

The Formal Professional Biography

A good CV or résumé relates your strengths and achievements to your purpose—professional review or job search. Your document needs to be drafted in the clearest language, with the most attractive and functional design (see Figures 18.1 and 18.2). Many personnel officers claim that résumés have less than one minute to make the right impression.

Build Content from Component Parts

Build your CV or résumé of component parts—modules that can be formed and re-formed to map your strengths and achievements. Within each module, present the most recent information first. Which elements should you use? In what order? Let your own achievements be your guide. For some occasions, you may want to emphasize details of academic training. In other cases, you may want to emphasize specialized skills, giving that module a more prominent position on the résumé. Match the concerns of your audience and the purpose of the document.

Curriculum Vitae

Jennifer N. Chau

XXXX El Colegio Road #xxx
Anytown, XX 00000
(000) 123-4567
Chau@xyz.edu

Education

University of California, Santa Barbara
B.S., Chemistry, March 2001
Relevant Courses

- Advanced Organic Chemistry
- Analytical Chemistry
- Biochemistry
- Diff. Eqn. & Linear Algebra
- Advanced Inorganic Chemistry
- Molecular Genetics
- Physical Chemistry
- Physics
- Physiology
- Spectroscopy
- Professional Science Writing
- 3-D Vectors Calculus

Research Experience

Department of Chemistry and Biochemistry, UCSB, 1999

- Optimized asymmetric methods for synthesis of chiral six-member rings.
- Synthesized anticancer, antiviral agents, Frondosin A and Sphingomyelinase Inhibitors.

Department of Molecular and Developmental Biology, UCSB, 1998

- Implanted cancerous cells with plasmid proteins.
- Examined chromosome movements using gene fusion.

Publications

"New Construction of Ortho Ring Alkylated Phenols via Generation and Reaction of Assorted Ortho Quinone Methides" R. W. Van De Water, D. J. Magdziak, Jennifer N. Chau, and T. R. R. Pettus. Journal of American Chemical Society (JACS) 2000, 122, 6502–6503. Available: http://208.209.231.30/cgi-bin/jtext?jacsat/122/i27/html/ja994209s.html.

Skill

Laboratory Techniques

- NMR, UV, & IR spectroscopy
- pH measurement
- Vacuum distillation
- HPLC
- Titration
- Recrystallization
- TLC and gas chromatography
- Sample extraction and preparation
- Cell splitting

Teaching Experience

California Achievement Minority Program (CAMP) facilitator in organic chemistry and calculus; funded by the National Institute of Health, 1999–2000

Languages

Fluent in Vietnamese

Selected Honors

- National Science Foundation Research Fellowship, 2000
- Ralph Kilb Summer Undergraduate Research Fellowship, 2000
- Distinction in Major of Chemistry, 2000

Membership

- American Chemical Society

Figure 18.1
In her curriculum vitae (CV), Jennifer N. Chau provides extensive information about her academic preparation as well as her research and teaching experience. (Used with permission.)

James P. Manassas

Home Address:
Anytown, XX 00000
(000) 123-4567

Office Address:
Anytown, XX 00000
(000) 123-4567
jmanassas@xyz.edu

Objective

Position in dynamic systems modeling and control.

Education

MASSACHUSETTS INSTITUTE OF TECHNOLOGY Cambridge, MA
Candidate for Ph.D. degree in Mechanical Engineering, June, 2001. Thesis under Professor P.N. Grant on "Lateral Dynamics and Control of Rail Vehicles."National Science Foundation Graduate Fellow, 1999–2001.

CORNELL UNIVERSITY Ithaca, NY
Bachelor of Science degree in Mechanical Engineering, June 1997. Broad curriculum in mechanical engineering with emphasis on mechanics, vibration, acoustics. Business administration courses included finance and organization development.

Experience

C.S. DRAPER LABORATORY Cambridge, MA
Research Assistant. Determined optimum resolution element for pilot warning indicator. Assisted in development of generalized aircraft simulation. *Summers, 2000 and 1999.*

PRINTYPE, INC. Maynard, MA
Research Intern. Developed, installed computer-controlled photo-typesetting system. *Summer 1998.*

DEERING-MILLIKEN CORPORATION Spartanburg, SC
Operation System Analyst. Modeled, optimized monitoring of corporate operations. Performed linear programming, statistical analysis, manufacturing studies. Developed new inventory control system. *September, 1996–August, 1998.*

Publications/ Presentations

Manassas, J.P., "Wheel Wobble on Uploaded Freight Cars," *Process and Control, Vol.15,* No. 3, May, 1999, pp. 142–145.

Grant, P.N. and Manassas, J.P., "Control of High Speed Rail Vehicles" presented at the Fifth International Process Control Conference, June 1999, Tokyo, Japan.

Figure 18.2
A résumé always highlights a professional objective, while it also provides information about education and relevant experience. (Modified and used with permission, MIT Office of Career Services and Pre-professional Advising.)

Some résumé elements are conventional, but only the first of the following is absolutely required:

- Name, address, telephone number, e-mail address
- Objective
- Educational history
- Employment history
- Skills and training
- Licenses and certifications
- Honors and awards
- Memberships
- Publications
- Conference presentations
- Personal background

Edit for Clarity and Focus

Audiences for résumés are knowledgeable, demanding, exceedingly critical, short of time, and anxious to deselect even marginally unsuitable résumés as quickly as possible. Experienced résumé readers quickly eliminate the apparently unsuitable and move on to those applicants most likely to survive further intensive review. Ask several colleagues to read your draft CV or résumé, looking for obscure presentation of information. You know what the U of O is, while your reader may wonder if it is the state university of Ohio, Oklahoma, or Oregon. Combine miscellaneous jobs into one category, and don't inadvertently emphasize the insignificant. Remember that the first audience for your résumé may be a computer—an electronic résumé management system that scans documents looking for keywords. To reach this nonhuman audience, include nouns and verbs that clearly specify your skills and accomplishments. Whether your first readers are humans or computers, help them to skim your résumé efficiently by using phrases rather than complete sentences. Consider presenting your technical accomplishments in the form of a Skills Matrix (see Figure 18.3), included in the body of the résumé or prepared as an attachment.

Design for Hard Copy and Electronic Delivery

In recent years, the content of the résumé or CV has remained relatively constant for technical professionals, but the appearance has changed.

SKILL	Proficiency: 1 = Novice 3 = Intermediate 5 = Expert
C++	1
Java	2
Javascript	3
HTML	5
XML	2

Figure 18.3
Many job applicants prepare a skills matrix to attach to the résumé. The skills matrix is a good vehicle for summarizing specialized knowledge and accounting for your level of preparation in each area.

Because of the widespread availability of desktop publishing software and laser printing, résumés and CVs look much more professional than they used to. If you have not updated the design of your résumé, it may look underprepared in comparison with others.

However, you have the challenging task of providing your recipient with a document that looks good in hard copy and also maintains its professional appearance when transmitted via e-mail, faxed, scanned, or posted on a Web site. In hard copy, deliberate and consistent use of page design elements like white space, bullets, italics, and bold type can ensure that your strengths are apparent even to readers who do not spend much time studying your résumé. But when your résumé is scanned or sent as an e-mail attachment, the same typographical features might hamper a computer's ability to "read" your résumé. To improve the chances that your potential employer will be able to scan your document, use high-quality, standard letter size white paper ($8\frac{1}{2} \times 11$ inches), print on one side only, and do not fold or staple. Be wary of formatting flourishes like bullets, which can sometimes morph into other symbols when your

CV of Jennifer Chan	Page 2 of 2

Figure 18.4
If your CV or résumé is longer than one page, use an identifying heading on all pages beyond the first.

résumé is read by a digital system. You can get at least some idea of how your résumé will hold up to e-mailing, scanning, downloading, uploading, and keyword-searching by sending your résumé to colleagues who can then let you know if your document maintains its formatting on their computer systems.

Streamline and Update

How long should a CV or résumé be? With hard copy résumés, many organizations are strongly committed to the one-page limit, even for senior professionals who could fill pages with their achievements. For others, a multiple-page document is acceptable. Electronic résumés are often longer than one page. Find out what you can about what is expected, but assume that you have limited space to work with. A CV or résumé should be as short as possible, and you will want to get the maximum payoff from every line. If you do exceed one page, be sure to put a heading on additional pages with your name and the page number (Figure 18.4).

Keep your CV current. It should reflect the changes in your life, and you should have a procedure for making changes, for dropping some data and adding others. Because a CV should be brief, you will need to weed out older achievements as you add more recent ones. As you move forward, detailing more recent job descriptions, you will take fewer steps backward, inevitably dropping one or more older pieces of your life to make room for the new.

Job-Related Correspondence

Cover Letters

A letter offering yourself for employment should be brief but detailed and informative. Except in unusual circumstances, it should be written *to* someone, and it should indicate that you are responding to an actual situation. If you want the job, do your best to find the name of an appropriate recipient. The letter of application in Figure 18.5 indicates that the writer has acquainted herself with the company's research agenda. She elaborates on the enclosed résumé, providing details that will interest the potential employer. The letter is friendly without exaggeration or insincerity. Even when you submit a résumé by way of e-mail, don't neglect a cover letter. In that case, name the software with which your résumé has been prepared.

Writing letters like this is hard work. It is certainly easier to write a brief note asking "To Whom It May Concern" to study your enclosed résumé and decide which of your achievements would match the company's needs. But the burden of connecting what you have to offer with what the employer requires is on you, the writer. You want something from your reader, and you should make it easy to see what distinguishes you from other applicants. If you are sending your letter via e-mail with a résumé attached, be sure to say which software and version you used to create your résumé. Some experienced job seekers send a hard-copy version of their application letters and résumés in addition to the e-mail version.

When you submit a hard-copy application, consider the envelope in which you enclose your letter and résumé. Because the envelope is postmarked and date-stamped, it frequently remains a part of your file. You can make the package more attractive by using an envelope large enough to accommodate your documents without folding and by designing a mailing label that matches your letterhead.

List of References

It is helpful to potential employers if you submit an annotated list of references with your letter of application and résumé, providing recently validated contact information as well as a brief indication of your connection with the person named (see Figure 18.6). However, you should get explicit permission from all persons on your list before you provide their names for this purpose. A reference letter from someone who does not want to write it is not much use. You can get permission by telephone, of course. But we think that you will increase your chances of getting a letter that focuses on your fitness for the job if you write to your referee, hard copy or e-mail, calling attention to aspects of your experience that match your potential employer's needs. Always enclose

Sarah Wilson • Baker House #332 • 258 Memorial Drive • Cambridge, MA 02139

13 November 2005

Ms. Maria Gonzales
IBM
Director of Personnel
20525 Mariani Avenue
Cupertino, CA 95014

Dear Ms Gonzales:

I will be graduating from MIT in June with a Bachelor's degree in Mechanical Engineering. From your company Web page, I see that IBM is actively involve with robotics and automation control systems, an area of engineering that particularly interests me. I have enclosed a copy of my resume to give you more information on my academic, research and industrial experience.

At MIT, I have taken coursework in computer-aided design, control systems manufacturing, and electronics, in addition to the general course requirement for mechanical engineering. For the past two summers I worked on designing disk drives at Digital Equipment Corporation. I prepared detailed engineering sketches and assisted with quality control. Given my academic training and workplace experience, I believe I can make a contribution to IBM and learn a great deal at the same time.

I would welcome the opportunity to meet with you and discuss employment opportunities at IBM. You can reach me by e-mail at swilson@mit.edu. I look forward to hearing from you

Sincerely,
Sarah Wilson
Sarah Wilson

Enclosure: Resume

Figure 18.5
The author of this letter of application to the IBM Director of Personnel has skillfully indicated her familiarity with IBM research activities. She then summarizes her relevant academic training and work experience. (Modified and used with permission, MIT Office of Career Services and Pre-professional Advising.)

Emma Y. Lee

E-Mail:Lee@chemistry.ucsb.edu

References

Archie Barnes. Lead Laboratory Assistant, Clinical Laboratory, Good Samaritan Hospital, San Jose, CA 95124. (408) 559-2474.

> Mr Barnes is my immediate superior at the clinical laboratory where I work. He may be reached from 7:30 a.m.–4:00 p.m.

Vojislav Srdanov, Ph.D. Associate Professor, Department of Chemistry, University of California, Santa Barbara, CA 93106. E-mail<srdanov@chem.ucsb.edu>

> I work as an assistant researcher for Dr Srdanov. He supervised my projects and trained me to operate ESR, RAMAN, NMR, and X-ray diffractometer. He can be reached through e-mail.

Fiona Goodchild, Ph.D. Education Coordinator, QUEST, University of California, Santa Barbara, CA 93106. E-mail<fiona@mrl.ucsb.edu>

> Dr. Goodchild was the coordinator for my three research internships in the Department of Chemical Engineering, Marine Biology, and Chemistry. She can best be reached through e-mail.

Figure 18.6
If you are asked for a list of references, be sure to provide enough information so that potential employers can easily contact each one. This job candidate has taken the time to indicate her professional connection with the referees. (Used with permission.)

an updated résumé. Inform your referees of the outcome of your job search and thank them for the time they have spent on your behalf.

Follow-Up Letters

Many job applicants write thank-you letters to companies where they have interviewed. Writing to thank someone for an interview can, in tight circumstances, make the difference between being hired and being second in line. It suggests that you'll be considerate and pleasant to work with. Declining an offer presents another important occasion for a letter. If you decide to say no, a courteous acknowledgment is certainly in order. Writing letters declining offers can also pay off, though sometimes years later. In the small networks that form any subspecialty, having the grace to say "no" formally may make you worth remembering when new projects come up.

CVs and Résumés in Transition

Several commercial services now disseminate electronic résumés (see, for example, ⟨www.monster.com⟩; ⟨www.hotjobs.com⟩). Some post online versions of conventional résumés, while others provide applicants with a résumé template to complete and also a selection of prewritten cover letters to modify with personal details.

Many energetic job seekers now design multimedia résumés to display on personal Web sites. These typically begin with a conventional résumé. Each element can be followed in depth, with amplification that includes text, sound, still picture, and moving picture. The education module might link to a list of all undergraduate courses taken, while the name of each course could link to a course description and syllabus. Possibilities are very broad and include full text of laboratory reports or other documents written in a course, or videos in which the applicant demonstrates expertise by performing specific tasks.

Though very few employers expect a multimedia résumé, it is easy to imagine a future in which the conventional, one-dimensional résumé will be obsolete. For now, however, an informative and well-designed résumé or CV will be useful in your professional life, particularly if you have ensured that it can be scanned by computers as well as by human readers.

A Brief Handbook of Style and Usage

1. *Write effective paragraphs.*
2. *Break long sentences into manageable units.*
3. *Make choppy writing flow.*
4. *Use parallel subject headings to reveal logical flow.*
5. *Emphasize the active voice.*
6. *Write with economy.*
7. *Avoid the abstract prose caused by excessive nominalizing.*
8. *Put parallel objects, actions, and thoughts into parallel sentence elements.*
9. *Don't line up long strings of modifiers in front of nouns.*
10. *Place modifiers close to the words they modify.*
11. *Make your pronouns refer clearly to the objects and ideas that they stand for.*
12. *Make words related by number, pronoun reference, and case agree with each other.*
13. *Use definite articles (the) and indefinite articles (a, an) to identify the status of nouns.*
14. *Use words carefully.*
15. *Don't use language that stereotypes or excludes other people.*
16. *Use commas to help the reader sort elements in the sentence.*
17. *Capitalize proper nouns, book and article titles, certain scientific terms, and references to chapters, equations, figures, and tables.*
18. *Use apostrophes to identify possessives, plurals, and contractions.*
19. *Use hyphens to form compound words and compound modifiers, and to divide words.*
20. *Use semicolons to join closely related clauses and to separate certain items in a series.*

21. Colons set off elements that amplify, explain, or illustrate the main clause.

22. A dash sets off material you want to emphasize.

23. Parentheses and brackets enclose added detail or commentary within the sentence.

24. Abbreviations and acronyms can save space and eliminate repetition.

25. Use figures rather than words for numbers in scientific and technical prose.

26. Italicize the titles of particular vessels, longer publications, words featured as words, and foreign words.

27. Use quotation marks to set off quotations and the titles of some short works.

You're about to leave for the day, but, glancing at your computer screen, you see a new message waiting. It's from a member of your group, so you open it. A file is attached. Your two colleagues want you to review that design study one final time. It's due first thing next week, and the client—a major source of business for your group—is known to be a stickler for deadlines, accurate detail, and clear writing. As you glance over the attachment, you begin to feel uncomfortable as you slow to a crawl in patches of wordiness, dense prose, and awkwardness. Your two colleagues are brilliant, but neither is a particularly fluent writer. You wonder if you can get someone from the office staff to look at the report tomorrow, but the last time you asked, they just ran the piece through a spell checker and added some capitals and commas. Your organization's editorial service takes a week to turn a priority document around. As you leave your office, you realize that you're the only one left to do the job. You need a refresher on style and usage.

Writers in the sciences and applied sciences place a high value on accurate description and detailed analysis. They use numbers, symbols, and special terminologies to treat subjects that range in scale from the subatomic to the cosmic. This quest for accuracy and detail in subjects that can be challenging to represent often leads to the problems in style and usage we associate with foggy, dense writing in the professions. Our handbook identifies some of the more common of these problems and recommends ways of avoiding or correcting them. We begin with some guidelines for writing paragraphs, continue with suggestions for sentences and word use, and close with some key points on punctuation.

1. Write effective paragraphs. Paragraphs are the main building blocks of writing. Constructed tightly, they advance your thoughts in clear stages. A well-structured paragraph is commonly arranged as a topic sentence, or main idea, and one or more supporting sentences that develop the idea by means of description, narration, exemplification, or analysis. The example below shows a pair of paragraphs with topic and supporting sentences that are linked together by a series of *keywords* (KW).

Wind Shear: Phenomenology of Microbursts and Gust Fronts

Downdrafts within storms generate microbursts and gust fronts [Topic sentence]. Through a variety of processes, including evaporative cooling and precipitation loading, negatively buoyant air within a storm descends to the ground as a downdraft [Keyword (KW)]. Upon reaching the surface, the downdraft [KW] produces a pool of cold air known as the outflow. Microbursts [KW] are formed when the divergent air beneath the downdraft [KW] reaches a specified intensity, namely an increase in wind speed of greater than 10 m/s over a distance of less than 4 km. These [Demon. pronoun] downdrafts [KW] pose hazards for landing aircraft [Transitional Sentence].

An aircraft's [KW] potentially hazardous encounter of a microburst [KW] is illustrated in Figure 1 [Topic sentence]. Upon entering the microburst [KW], the aircraft [KW] first experiences an increase in headwind. This [Demon. pronoun] increase causes the aircraft [KW] to fly above the glide slope. The pilot, who is often unaware of the microburst [KW], may attempt to return to the angle of attack. As the aircraft [KW] continues through the microburst [KW], it encounters a strong downdraft [KW] and then a tail wind, which results in a loss of lift. The aircraft [KW] falls beneath the glide slope and the pilot must now increase power and angle of attack to bring the aircraft [KW] back to the glide slope. The aircraft [KW], which requires a finite amount of time to respond to the controls, crashes if it is too close to the ground to recover.

You can make paragraphs like the two above hold together by means of three organizing principles:

(1) *Unity.* Focus on a topic that will unify the content of the paragraph. Do not shift to new topics in mid-paragraph. All the sentences of the paragraphs above contribute to the two main topics, represented by the keywords *microbursts* and *aircraft*, which are woven throughout the paragraph.

(2) *Development.* Advance the topic by means of some expository strategy such as description, narrative, exemplification, definition, comparison and contrast, or analysis. Notice that the two paragraphs in the example above develop through a combination of description and analysis.

(3) *Coherence.* Make the paragraph sentences hang together through various linking strategies, including keywords, demonstrative pronouns, and transitions. Some of these strategies for linking the sentences together in the two paragraphs above are identified in brackets.

2. Break long sentences into manageable units. Long, dense sentences, often amounting to more than 30 words, may contain more information than a competent reader can readily understand. You can usually identify these sentences by their awkward clause structures. Determine the main actions of the sentence. Then sort these into two or more shorter sentences.

Long, awkward sentence

In gasoline engines, designers leave a space between the piston and its cylinder that contributes to the exhaust emission problem, because as the engine is started and begins to heat up, the cylinder liner, which is directly cooled by a surrounding coolant, expands more slowly than the piston, which allows exhaust gases to escape. [Length: 54 words. This sentence contains too many clauses—and ideas—for easy processing.]

Improved

Gasoline engine designers leave a space between the piston and its cylinder that contributes to the exhaust emission problem. At startup, when the engine begins to heat up, this space allows the cylinder to expand rapidly without damaging the more slowly expanding cylinder liner, which is directly cooled by a surrounding coolant. The space, however, also allows exhaust gases to escape. [Length: 61 words.]

Although it took eight more words to write this second version, its three sentences develop the information with greater clarity.]

3. Make choppy writing flow. Choppy sentences interrupt the smooth flow of thought, and they can be repetitious. Combine overly short sentences with the help of transitional words, coordinating conjunctions (e.g., *and*, *yet*, *but*, *nor*, *or*), and subordinating conjunctions (e.g., *unless*, *since*, *because*, *if*, *when*).

Choppy writing

Cytolytic toxins act directly on cell membranes. They disturb the normal physiology of the target cell. They ultimately kill the cell. Cytolytic toxins are not a single group of related chemicals. They are not produced by one class of organism. These toxins are heterogeneous in their chemical structures. They can be obtained from plant and animal sources. They do not share a common mechanism of action. There are several ways for cytolytic toxins to interfere with the normal permeability barrier formed by the cell membrane. See Table 1.

Improved

Cytolytic toxins act directly on cell membranes by disturbing the normal physiology of the target cell and [Coord. Conj.] ultimately killing it. These toxins are not a single group of related chemicals produced by one class of organism. Rather [transitional word], they are very heterogeneous in their chemical structures and [Coord. Conj.] can be obtained from both plant and animal sources. Consequently [transitional word], cytolytic toxins do not share a common mechanism of action, but [Coord. Conj.] have several ways, as summarized in Table 1, of interfering with the normal permeability barrier of the cell membrane.

4. Use parallel subject headings to reveal logical flow. Technical subject matter sometimes is so dense with terminology and operations that even well-designed paragraphs are difficult to follow. The reader must struggle to work out the natural hierarchy of ideas. Subject headings often help by marking out topical patterns of subordination and parallelism in otherwise opaque prose. In the example below, the revised version communicates at a glimpse the essential logic, and this explicit structure in turn enables the reader to get information out of the paragraph more effectively.

Dense prose

3.4 Unresolved Issue Number 4. The criteria for restarting Facility XYZ have not been met for the water drain capacity of the filter compartment, the stability of the charcoal in the absorber, and the capacity of the absorber.

The above unresolved issue consists of three separate restart criteria. The details of these criteria are as follows. The first criterion is that filter compartment water drains shall be demonstrated to be capable of meeting their design function. The second criterion is that the possible iodine desorption and autoignition that may result from radioactivity-induced heat in the carbon beds shall be considered when determining the adequacy of the charcoal absorbers. Finally, the third criterion is that the absorber section of the XYZ facility shall contain impregnated activated carbon filters demonstrated to remove gaseous iodine from influent. The carbon filters must have an average atmosphere residence time of 0.25 seconds per 2 inches of absorbent bed. The maximum loading capacity ... The capacity of the water drains (the first criterion) is addressed in RRD-RSE-910003, "Revi-sion to Filter Compartment Drain Capacity" (21 January 1991) ...

Revised version, with headings showing subordination and parallelism

3.4 Unresolved Issue Number 4

The criteria for restarting Facility XYZ have not been met for the water drain capacity of the filter compartment, the stability of the charcoal in the absorber, and the capacity of the absorber.

3.4.1 Criteria for Restarting

The following criteria must be met before Facility XYZ may resume operations:

3.4.1.1 Capacity of filter compartment drain lines. The filter compartment water drains must be demonstrated to be large enough to handle the capacity called for in the design.

3.4.1.2 Stability of the absorber's charcoal bed. The absorber's charcoal bed must be shown to be stable enough to prevent any possible iodine desorption and the autoignition that might result from radioactivity-induced heat in the carbon beds.

3.4.1.3 Use of the carbon filters in the absorber section. The absorber section shall use impregnated activated carbon demonstrated to remove ...

3.4.2 Assessments and Conclusions
The above criteria may be met as follows:
3.4.2.1 Capacity of the filter compartment drain lines.
Guidelines for regulating the capacity of the water drains are addressed in "Revision to Filter Compartment Drain Capacity" (RRD-RSE-910003, 21 Jan 1991) . . .

*5. **Emphasize the active voice.*** Although the passive voice has many legitimate uses, overusing it can lead to indirect, wordy prose. The passive voice inverts the straight *agent-action-thing acted upon* (i.e., *subject-verb-direct object*) sequence of the sentence. The thing acted upon becomes the subject of the sentence. For example, *Enzymes break down proteins* becomes *Proteins are broken down by enzymes*. Both sentences are grammatically correct, but the active verb *break down* is more direct and simple than the passive verb *is broken down*. The word order of the direct sentence is easier to process.

Passive/indirect writing

Different types of protein are broken down by different enzymes, and starch is dismantled by still other enzymes into its constituent sugar molecules.

Active/direct writing

Different enzymes break down different types of protein, and still other enzymes dismantle starch into its constituent sugar molecules.

The question of using the passive voice is often a matter of emphasis. The writer of the sentence above who is discussing proteins and wants to keep *protein* as the subject will choose the passive form. The writer who wants to maintain sentence focus on enzymes will choose the active form, which makes enzymes the subject.

Here are some instances in which the passive voice leads to awkward, wordy, or ambiguous expressions:

Awkward passive

A heat barrier installation has been carried out by the plant maintenance crew.

Improved

The plant maintenance crew installed a heat barrier.

Wordy passive

The cost of the filtration system was found by the research team to be justified, because a greater efficiency in the performance of the instrument was obtained. [27 words]

Improved

The research team justified the cost of the filtration system with the instrument's greater efficiency. [15 words]

Ambiguous passive

Sensing information must be provided manually when the device is in the manual mode. [Not clear who is doing what here.]

Improved

The shift operator must provide sensing information manually when the device is in the manual mode.

6. Write with economy. Writers often draft wordy, convoluted prose, which needs to be condensed. Cutting unnecessary words and phrases improves the clarity and impact of your writing.

Wordy writing

The cooling of the thermal unit is accomplished by using electric fans which are run every other hour during the day. [The "empty" verbs accomplish and run may be eliminated without loss of meaning.]

Improved

The thermal unit is cooled with electric fans every other hour during the day.

Wordy writing

An increase in water volume would have the effect of reducing the stability of the slope along the North wall of the power plant. [... have the effect of ... contributes nothing to the sentence meaning.]

Improved

Increased water volume would reduce slope stability along the power plant's North wall.

Wordy writing

There was a secondary stress that was identified with the stress caused by constrained thermal expansion of the pipe fitting. [Avoid empty clauses like there is or there was at the start of sentences. Note also the repetitious use of stress.]

Improved

A secondary stress was caused by constrained thermal expansion of the pipe fitting.

Certain phrases show up repeatedly in wordy writing. Here are some of them:

Wordy	Direct
... at the present time ...	now
... due to the fact that ...	because
... has the capacity to ...	can
... have the effect of ...	—
... in the event that ...	if
... in the neighborhood of	about
... it should be noted that ...	note that
... has been conducting an analysis of ...	has been analyzing

Wordiness sometimes originates in words and phrases that repeat what has already been stated or implied in the sentence. Compound formations (nouns, verbs, adverbs, and adjectives), for example, are a common source of repetition.

Repetitious verbs

Ring currents were observed and demonstrated to play a role in fullerene magnetism.
[*Improved*: ... were demonstrated to play ...]

Repetitious sentence complements

Mouse and human receptors are so different and distinct that ...
[*Improved* ... are so different that ...]

Repetitious ideas

The main cost of the hydro unit is determined by the costs of the catalyst and the frequency of its replacement. Catalyst life also is the major factor on the overall economics of operating the hydro unit.
[Both sentences are noting that replacing the catalyst is the main operating cost of the unit in question.]

Improved

Catalyst life largely determines the economics of the hydro unit, because its main cost is catalyst replacement.

Redundant words

... was a close approximation to ... [... was similar to ...]
... with absolute certainty ... [... with certainty ...]
... was blue in color ... [... was blue ...]
... round in shape ... [... round ...]

7. ***Avoid the abstract prose caused by excessive nominalizing.*** *Nominalizing* means forming nouns from verbs. You take a verb like *detect*, change it to an abstract noun like *detection*, and add a passive general-purpose verb like *has been achieved*. From these changes you can get a sentence like the following:

Wordy nominalization

The detection [Abstract noun made from verb *to detect*] of intracellular products of polymerase chain reactions has been achieved [Passive general-purpose verb] by two very different methods.

This example may be simplified by restoring the main action, *detect*, to the verb position of the sentence:

Improved

Intracellular products of polymerase chain reactions have been detected by two very different methods. [The main action has been restored to the verb.]

Alternate improvement

We have detected intracellular products of polymerase chain reactions by two very different methods. [By restoring the first-person agent, we get an active verb.]

Technical prose uses a lot of nouns like *detection* as a way of focusing on abstract concepts or processes. Yet, nominalized words can produce awkward, wordy writing, with lots of abstract nouns supported by empty general-purpose verbs. Here is another example:

Wordy nominalization

Measurement of the levels of about 6,800 different genes in bone marrow samples was carried out on 38 leukemia patients.

Improved

Bone marrow samples of 38 leukemia patients were measured for levels of about 6,800 different genes.

8. Put parallel objects, actions, and thoughts into parallel sentence elements. Sentence parallelism is the practice of arranging similar ideas into coordinate patterns of verbs, nouns, phrases, or clauses. These patterns emphasize the similarities of the ideas and help make the writing clear. Faulty parallelism creates muddy sentences that require careful rereading to sort out. Here are some of the sources of faulty parallelism:

Faulty parallelism in a series

Microscopic mechanical systems (MEMS) can respond to a variety of inputs, including light, heat, and vibrating objects that cause stimulations. [The end series of the sentence includes two nouns, *light* and *heat*, followed by a clause.]

Improved

Microscopic mechanical systems (MEMS) can respond to a variety of inputs, including light, heat, and vibrations.

Faulty parallelism in larger sentence elements

The negative Doppler effect increases both because of the increasing fraction of resonance absorber [phrase] and the neutron energy spectrum is lowered [clause]. [A phrase is mismatched with a clause.]

Improved

The negative Doppler effect increases both because the fraction of resonance absorber is increased and the neutron energy spectrum is lowered. [Two clauses balance the sentence.]

Faulty parallelism also develops in incomplete constructions.

Faulty parallelism from an incomplete construction

Zinc exerts a greater effect on the vulcanization of isoprene rubbers than synthetic rubbers such as SBR and BR. [The sentence is comparing zinc to synthetic rubbers. The author intended to compare zinc's effect on isoprene rubbers with its effect on synthetic rubbers.]

Improved

Zinc exerts a greater effect on the vulcanization of isoprene rubbers than it does on the vulcanization of synthetic rubbers such as SBR and BR. [The construction setup in the first part of the sentence has been completed.]

Sentence parallelism also helps keep ideas clear. The sentence below is grammatically correct, but the two clauses arrange the parallel items in very different ways. The result is hard to read:

Faulty parallelism in ideas

Under a centrifugal stress exceeding 5000 psig, the test alloy fractured along the weld seam of surface A; the rupture along the horizontal axis of Surface H of the alloy occurred under a hydrostatic pressure that exceeded 10,000 psig. [The subjects and verbs of the two independent clauses do not contain parallel information. The subject of the first half of the sentence is *alloy*; the subject of the second half of the sentence is *rupture*.]

Improved

Under a centrifugal stress exceeding 5000 psig, the test alloy fractured along the weld seam of Surface A; under a hydrostatic pressure exceeding 10,000 psig, the alloy ruptured along the horizontal axis of Surface H. [The two subjects and verbs contain parallel information, which makes the improved version easier to follow.]

9. Don't line up long strings of modifiers in front of nouns. Science and technical prose depends heavily on modification for achieving its accuracy. In an effort to be accurate, writers often stack up modifiers in front of the main noun. The true effect of these modifier "stacks," however, is not accuracy but ambiguity. The reader has to work out which words are modifying other words in the stack. For example, in "underground plant effluent soil contamination," the adjective *underground* could be modifying either *plant* or *contamination*. The phrase could be referring either to "contamination from an underground plant" or to "underground contamination from an above-ground plant." To resolve this ambiguity, we put some of the modifying information after the main noun: "underground soil contamination by a plant effluent." Here are some additional examples of stacked modifiers:

Stacked modifier

Large low-cost central receiver electricity generating power plants could significantly alter local desert climates by modifying their radiation balances. [The main subject, *plants*, is modified by 8 preceding words. Do the adjectives *large* and *low-cost*, for example, apply to *receiver* or to *plants*?]

Improved

Large electricity-generating power plants of the low-cost central receiver type could significantly alter local desert climates by modifying their radiation balances. [Some of the modifiers have been shifted to the phrase that follows the main subject.]

Stacked modifier

A contributing cause of the accident was the poor communication among health protection and environmental safety group personnel and operations management. [The 6 modifiers in front of personnel make it hard to tell how many groups are implicated in this sentence.]

Improved

A contributing cause of the accident was the poor communication among the personnel of the health protection group, the environmental safety group, and operations management. [*3 groups*]

Alternate improvement

A contributing cause of the accident was the poor communication between the personnel of the health protection and environmental safety group and operations management. [2 *groups*]

Stacked modifiers are common in technical titles.

Stacked modifier in a technical title

An Interdisciplinary Study of Coupled Atmosphere-Ocean Model Circulation Flux Adjustments [What is being modeled?]

Improved

An Interdisciplinary Study of Flux Adjustments in Circulation Models of Coupled Atmosphere-Ocean Systems.

Alternate improvement

Flux Adjustments in Circulation Models of Coupled Atmosphere-Ocean Systems: An Interdisciplinary Study [A two-part title]

10. Place modifiers close to the words they modify. A modifier becomes ambiguous when it is not closely linked to the item it is modifying.

Don't put modifying words and phrases into out-of-the-way places in the sentence.

Misplaced modifying phrase

The storage drums showed signs of deterioration that could be seen under severe corrosion. [The phrase *under severe corrosion* appears to be modifying *seen* rather than *drums*.]

Improved

The storage drums, which were severely corroding, showed visible signs of deterioration.

In dangling modifiers, a common form of the misplaced modifier, a word or phrase modifies a noun that is not the target. In the sentence "Walking down the street, the tall buildings came into view," the writer is suggesting that the tall buildings are out for a walk. Although we can usually understand what a dangling modifier is trying to modify, danglers are errors of logic.

Dangling modifier

By carefully adjusting the reflecting surface spacing, the desired transmission wavelength can be isolated. [The action of adjusting the spacing is misattributed to *wavelength*, which is the subject of the main clause.]

Improved

By carefully adjusting the reflecting surface spacing, we can isolate the desired transmission wavelength. [The action of adjusting the surface is now attributed to the actual agent, *we*.]

Alternate improvement

Carefully adjusting the reflecting surface spacing [gerund phrase] will isolate the desired transmission wavelength. [Making a gerund phrase the subject of the sentence also eliminates the dangling modifier.]

11. Make your pronouns refer clearly to the objects and ideas that they stand for. Pronouns (e.g., *he*, *she*, *it*, *they*, *this*) refer back to a preceding noun (i.e., the referent). They help tie the different sections of the sentence or paragraph together without repetitiously mentioning the noun. It is easy, however, for a writer to think his or her pronoun is referring to something definite when, in fact, the referent is unclear.

Vague pronoun reference

Although the photosynthetic bacteria all possess structural bodies with localized photochemical apparatuses, their morphologies vary from species to species. [Does *their* refer to *apparatuses*, *elements*, or *bacteria*?]

Improved

Although the photosynthetic bacteria all possess structural bodies with localized photochemical apparatuses, the morphologies of their structural bodies vary from species to species.

A vague pronoun reference often forms when a pronoun that refers to a noun in the previous sentence appears alone as the subject of the next sentence.

Vague pronoun reference at the start of a new sentence

One of the distinctive features of Mars' southern hemisphere, the giant Hellas impact basin, is thought to have been formed from an asteroid hit. It is surrounded by a ring of ejected material that is more than a mile thick and reaches 2.500 miles from its center. [*It* and *its* could refer back to several nouns, including *features*, *hemisphere*, *basin* or *hit*.]

Improved

One of the distinctive feature of Mars' southern hemisphere, the giant Hellas impact basin, is thought to have been formed from an asteroid hit. The basin is surrounded by a ring of ejected material that is more than a mile thick and reaches 2,500 miles from the basin center. [The pronoun has been replaced with the noun that was its intended referent.]

12. Make words related by number, pronoun reference, and case agree with each other. A plural subject requires a plural verb (subject-verb agreement), a plural noun-referent requires a plural pronoun (pronoun-referent agreement), and a pronoun must agree with the case (case agreement) in which it is used.

Subject-verb nonagreement

The mixture of methanol and water used in the process were then recovered and distilled for further recycling. [The subject is *mixture*, which is singular and takes a singular verb.]

Improved

The mixture of methanol and water used in the process was then recovered and distilled for further recycling.

Collective nouns such as *committee* and *team* are treated as singular:

Subject-verb nonagreement

The five-nation Interstate Council for the Aural Sea have called for an increased cubic kilometer flow of water into the Aural basin. [*Council* is treated as a singular noun and should take a singular verb.]

Improved

The five-nation Interstate Council for the Aural Sea has called for an increased cubic kilometer flow of water into the Aural basin.

It is easy to mistake number agreement when the first part of your sentence uses a pronoun for its subject and is followed by a modifying phrase.

Pronoun-verb nonagreement

Each of the casings are constructed from 9 percent nickel steel, because they must withstand constant temperatures as low as −320 °F. [*Each*, the subject of the sentence, is singular and requires a singular verb, *is constructed*. The second pronoun, *they*, also does not agree with its referent, *each*.]

Improved

The casings are constructed from 9 percent nickel steel, because *they* must withstand constant temperatures as low as −320 °F. [The two pronouns are made into plurals.]

Nonagreement in pronoun case is less common in the writing of science and engineering than it is in college themes, but it does occasionally happen.

Pronoun case nonagreement

The responsibilities of laboratory management have been shifted to Roberts and I. [The pronoun is the object of the preposition *to* and should be in the objective case: *me*.]

Improved

... shifted to Roberts and me.

Pronoun case nonagreement

Roberts and him have assumed laboratory management responsibilities. [The pronoun is part of the subject and should be formed In the subjective case: *he*.]

Improved

Roberts and he have assumed . . .

13. Use definite articles (the) and indefinite articles (a, an) to identify the status of nouns. Articles may be a special problem if you are not a native English speaker, and your native language doesn't use articles. Here are some general guidelines:

(1) *Indefinite articles.* Use an indefinite article *a* (*an* before a vowel sound) for a singular count noun that has no special status. It may be an entity that is first being mentioned or it may be one of many similar entities. Count nouns are items that may be counted, such as *molecules* and *specimens*.

Indefinite article

We evaluated *a* group of five patients with auditory verbal hallucinations for . . . [The indefinite article tells us that this is the first time the group is being mentioned.]

DNA is a chain of double-stranded nucleic acid. [DNA is one of many such entities.]

Note that indefinite articles are used only with singular nouns and count nouns.

(2) *Definite articles.* The definite article *the* signals that the noun is a particular known entity. This particularity may be a result of the noun's having been mentioned in a previous sentence. It may also be that the noun has been specified as unique by means of added detail.

Indefinite article followed by definite articles

A zoom system employed on a metallograph gives a wide choice of magnifications. The zoom is fitted with a simple sheet-film holder . . . The film holder allows the sheet film to be manipulated on the metallograph by . . . [Note that *zoom system*, *metallograph*, and *film holder* all use the indefinite article when they are first mentioned. Once they have been introduced, we switch to the definite article.]

Definite articles used to signal particular entities

The catalyst, XR26, will increase the reaction rate in the hydro unit by 20 percent.

Avoid using the definite article for generalizations.

Generalization without a definite article

Microrobots can be made to work without central control from a computer. [The author is speaking about microrobots in general.]

Specific instance using the definite article

The microrobots can be made to work without control from a computer. [The definite article tells us that the author is speaking about a particular group of known microrobots.]

14. Use words carefully. Poor word choice is a major cause of cloudy prose. Four common kinds of poor word choice are: inaccurate words, vague words or phrases, abstract or ornate words, and awkward words.

(1) *Inaccurate words.* Think carefully about word meanings. Inaccurately used words (bad diction) are one of the most common of all errors in writing. Make a habit of using a dictionary to review the meaning of a word.

Inaccurate word use

Where widely divergent temperatures are expected ... [The author did not intend to suggest that the temperatures are expected to become increasingly different from some initial level.]

Improved

Where widely different temperatures are expected ...

Inaccurate word use

Variations in the types of wastes made it impossible to anticipate which cleaning solution would prove effective. [Suggests that the wastes themselves are undergoing changes, rather than that different kinds of wastes are being treated.]

Improved

Differences among the types of wastes ...

Inaccurate word use

The storm and subsequent high winds incurred heavy damage on the transmission equipment. [*Incurred* means "as the result of one's own actions" and does not apply here.]

Improved

The storm and subsequent high winds <u>inflicted</u> heavy damage ...
[*Inflicted* means "to cause by violent action."]

(2) *Vague words and phrases.* Be specific. Vagueness in writing also has its origins in word use. Some of your key words may mean a lot more to you than they do to the reader. Consider either replacing or modifying words and phrases that are not specific enough to give a clear and detailed impression.

Vague usage: ... a <u>moderate amount</u> of potassium permanganate solution ...
Improved: ... 10 ml of dilute potassium permanganate ...

Vague usage: ... the system will be exposed to <u>high temperatures</u> ...
Improved: ... the system will be exposed to temperatures of 400–450 °C

(3) *Ornate words.* Speak plainly. Don't choose words because they sound impressive. Science and technical prose will always require plenty of multisyllabic terms. Temper this complexity by using the simplest word or phrase that will do the job.

Ornate usage: The <u>linear integrity</u> of the pipe was compromised.
Improved: The pipe was bent.

We often associate ornate word use with the language of the highly educated. Here are some common sources of ornate usage:

Ornate	Simple
procure	get
become cognizant of	learn
efficacious	effective
endeavor	try
utilize	use
proactive	preventive
terminate	end
remuneration	payment
subsequent to	after
commence	begin
initiate	start

(4) *Awkward words.* Sentence awkwardness, the unskillful use of words, is perhaps the most persistent of all stylistic problems. It is generally the result of haste. Most drafts have some clumsy writing. Writers are concentrating on getting thoughts on the page and may not be choosing their words very carefully. This problem can only be dealt with if you patiently read over your drafts and recast clumsy usages.

Awkward word usage:... are related to the penicillins both structurally and activity-wise ... [an awkward colloquialism]
Improved:... are related to the penicillins in structure and mode of action.

Awkward word usage:... an atomic and molecular level understanding of the dynamics of ... [*Atomic* and *molecular* are awkwardly combined with *level* as compound modifiers.]
Improved:... an understanding at the atomic and molecular levels of the dynamics of ...

15. Don't use language that stereotypes or excludes other people. Be sensitive to bias that excludes people from their organizational or social rights. When you use a word like "chairman" in a generalization, you imply that only men are fit for such a role. Here are some other examples of bias in writing:

Biased role reference: The departmental chairmen will meet each month to review the progress of ...
Improved: The departmental chairs will meet each month to review ...

Gender-specific pronouns (*he*, *she*, *him*, *her*) may also be sources of bias.

Biased pronoun reference: Each maintenance crew member is responsible for entering his own completed work orders in the maintenance database.
Improved: Maintenance crew members are responsible for entering their own ...
Alternate improvement: Each maintenance crew member is responsible for entering his or her own ...

Even the occasional verb may suggest bias:

Biased verb: The submersible vehicle will be manned by two ...
Improved: The submersible vehicle will be operated by two ...

Some common *man*-words associated with gender bias include: *manly*, *manhunt*, *man-hole*, *man-hour*, *man in the street*, *mankind*, *manpower*.

16. Use commas to help the reader sort elements in the sentence. The comma, one of the essential elements of punctuation, separates units of thought according to their functions in the sentence. Without commas, the reader misses many of the subtle shifts in thinking and often must reread the sentence in order to make sense of it. Here are the most common uses of commas:

(1) *After phrases and subordinate clauses that introduce the main clause.*

> Although heavy metals in a spill bind to the soil, fluorides and chlorides may still migrate to the ground water. [After introductory clause]
>
> In all 8 monkeys, the cuneate fasciculus of the spinal cord had almost totally disappeared as the result of the degenerating central axons. [After introductory phrase]

(2) *Before the coordinating conjunction*—and, but, for, nor, or, so, yet—*that connects two clauses:*

> A work order was issued in November 1997, but the work was not begun until February 2000.

(3) *To set off nonrestrictive (parenthetical) sentence elements:*

> The low levels of beta-gamma activity, which were detected at the storm sewer outfall N3, had maximum radionuclide concentrations of about half the derived concentration guide. [The commas signal that the clause in the middle of the sentence is *nonrestrictive*—i.e., parenthetical to the meaning of the sentence.]
>
> The low levels of beta-gamma activity that were detected at the storm sewer outfall N3 had maximum radionuclide concentrations of . . . [The clause is now *restrictive*, with no commas, making its information essential to the meaning of the sentence.]

(4) *Between words, phrases and clauses in a series:*

> If the tested material does not meet the design requirements, then select a new material, develop plans for periodic maintenance, or develop plans for periodic pipe replacement.

(5) *Between coordinate adjectives that could be separated with an* and:

> . . . a low-frequency, large-scale climate event . . .

17. Capitalize proper nouns, book and article titles, certain scientific terms, and references to chapters, equations, figures, and tables. Capitalization of a word is extremely helpful to the reader, because it conveys important information about the word's special status. In addition to signaling the beginning of a new sentence, capitals are used in the following:

(1) *Proper nouns and names*

Dr. Frank James [proper names, titles]
Friday, December [days of the week, months of the year]
the Missouri River, the Bering Sea [names of rivers, lakes, oceans]
Tucson, Minnesota, the Southern Hemisphere, Mount St. Helens, the Arctic, Mars, [names of places, geographical areas, planets]
World Ocean Circulation Experiment, Twenty-Fifth Annual Conference on Fluid Dynamics [specific events and undertakings]
the *Titanic, Landsat* 4 [vessels and vehicles]
the Soft X-Ray Telescope, the Rectilinear Heavy-Ion Collider [a unique, named instrument]
the Endangered Species Act [names of specific acts and laws]
the Baylor Instrument Company, Argonne National Laboratory, the Division of Organic Coatings and Plastics Chemistry, [corporations, institutes, and their units]
Rutherford backscattering [terms derived from a specific person's name]

(2) *Titles of works.* Practices for capitalizing titles will vary from field to field. Check with an editor or journal style guide.

Thin Film Processes [book title]
A Handbook of Case Histories in Failure Analysis [book title]
Journal of Experimental Medicine [periodical title]
Activation of Heat-Shock Genes in Eukaroytes [article title]
American Petroleum Institute Specification 5A [short document]
NACE Standard MR017S [short document]

(3) *Figures, tables, chapters in books and sections of documents.*

Figure 2, Equation 3-4, Chapter 6, Table 12-3
Figure 2. Fracture surface of one of the deeper fatigue cracks [figure title]
Chemical Probe Methods [section title in an article]

(4) *Acronyms and certain abbreviations and scientific terms.*

K [Kelvin], F [Fahrenheit], J [joule]: [Abbreviations for certain units of measure]
HIV, TB, NASA [acronyms]
Mycobacterium tuberculosis [genus-species terms]
CO_2, DNA [chemical symbols]

18. Use apostrophes to identify possessives, plurals, and contractions. The versatility of the apostrophe can make it a confusing punctuation mark for many writers. Sometimes we have to think carefully to decide whether an *-'s* formation is a possessive, a plural, or a contraction. Writers tend to confuse the functions of the apostrophe. It is easy to confuse the contraction *it's* (*it* + *is*) with the possessive *its*.

(1) *Possessive case of nouns.* Add *-'s* to singular or plural nouns not ending in *-s*. (Note the exception of the *its*, which is the possessive of *it* but takes no apostrophe.)

the electron's energy level [also: "the energy level of the electron"]
Parkinson's disease, Young's modulus [possessives formed with a proper name]
Saturn's main ring system
the medical society's meeting
the DOE's Office of Power Technologies [acronym treated as a singular possessive noun]
the virus's shape, octopus's eye [singular nouns ending in *-s*]
tuberculosis' typical development [Drop the *s* if the word sounds awkward.]

If the plural noun ends in *-s*, add the apostrophe only.

the researchers' results
nurses' health study
the emissions' CO_2 content

(2) *Plurals of numbers, abbreviations and letters.* Apostrophes are optional for many plurals, but the trend is to leave the apostrophe out. For example, the plural of the acronym *CPU* may be *CPUs* or *CPU's*. Choose one form and stick with it. Always use an apostrophe with a lowercase letter and an abbreviation.

two Ys, two Y's [plural of a upper case letter]
two 7s, two 7's [plural of a number]
two y's, 2 kg's [plurals of lowercase letters]
2 clm.'s (two columns), 2 mer.'s (two meridians), two meq.'s (two milliequivalents) [plurals of abbreviations]
the 1990's, the 1990s

(3) *Contractions.* Apostrophes may be used to combine some words.

it's [*it* + *is*. Use *its* for possessive of *it*.]
haven't [*have* + *not*]

19. Use hyphens to form compound words and compound modifiers, and to divide words. The hyphen is widely used to indicate word breaks at the ends of lines of print, but this process is automatic with most word-processing packages. The two most troublesome hyphen usages are those of compound words like the verb *cross-fertilize* and compound modifiers like *12-stage*. You can look up a compound word like *cross-fertilize* in a dictionary, where it will appear with the hyphen. Compound modifiers, in contrast, are coined by the writer for situations in which two words act as a unit to modify a third word. Compound modifiers do not appear as hyphenated words in the dictionary.

(1) *Compound words.* Avoid hyphenating two words that are merely associated with each other. "Electromagnetic wave" is not a compound word. Check the dictionary.

cross-fertilizing the specimens [verb]
the cable take-up [noun]
a decay half-life of 6 seconds [noun]
a sex-linked trait [adjective: a compound word that is also a compound modifier]

(2) *Compound modifiers.* Use a hyphen to connect words that work as a single unit of modification.

a 12-stage process
The Lorenz-Fitzgerald contraction, Sprague-Dawley rats
CO_2-capture process
mid-Atlantic [adjective formed with a prefix and a proper name]
an ac-to-dc converter [3-word modifier]
near- and long-term R and D objectives [two compound modifiers sharing the base word, *term*]

a low-frequency, large-scale climate event [two compound, coordinate modifiers, separated by a comma]

(3) *Numbers.* Hyphenate spelled out numbers from *twenty-one* to *ninety-nine.*

(4) *Other uses.* Hyphens also substitute for the words *to* and *through.*

1–2 kW at 27 MHz [as substitute for *to*]
0.1–1% dissolved solids
specimens 7–15 [as substitute for *through*]

20. Use semicolons to join closely related clauses and to separate certain items in a series.

(1) *To join two closely related independent clauses into a single sentence:*

The subjects' average total cholesterol levels fell by 9% on the walnut diet; the control Mediterranean diet led to a smaller decline of 5%.

(2) *To separate items in a series that already has commas:*

All CO_2-capture studies were based on commercially available equipment; all assumed effective controls on emissions of nitrogen oxide, sulfur oxides, and particulates; and all included the cost of compressing the CO_2 for pipeline transportation.

21. Colons set off elements that amplify, explain, or illustrate the main clause.

The most permeable polymers are the highly amorphous, glassy forms: their chain packings are sufficiently poor to permit penetrant access.

The three causes of cracks at the interface are:
- shrinkage caused by chemical activity
- shrinkage caused by drying
- expansion caused by hydration

[The colon preceding a set-off list is optional.]

We begin our study with Smith's concluding question: "Do the component wavelengths of light in the action spectrum have different effects on solar exposure?" [The quoted sentence begins with a capital.]

22. A dash sets off material you want to emphasize. The dash has two uses, but should be used sparingly. It can set off material for emphasis at the end of the sentence. In this role, it is similar to the colon, but has a more dramatic and informal effect. It can also be used in pairs in mid-sentence to set off parenthetical material. Indicate a dash in manuscript with two hyphens (--, typeset as an "em" dash, —).

A DVD audio sound recording will sample sound at extremely high frequencies—from 48,000 to 192,000 times per second. [at the end of the sentence]

The DOE estimates that between 1988–2020 most commercial low-level radioactive waste produced in the U.S.—80% of the volume and 97% of the radioactivity—will come from nuclear power reactors. [parenthetical comment]

23. Parentheses and brackets enclose added detail or commentary within the sentence. Note first that overuse of these devices may overload sentences with distracting detail. Parentheses and brackets, like a pair of commas and a pair of dashes, may also enclose supplementary material in the sentence. The comma pair is the standard and cleanest approach to including such detail, because the material fits syntactically within the flow of the sentence.

(1) *Parentheses.* Parentheses are effective ways to add a detail that does not merit being worked into the sentence.

An important role of the microscope's selected area diffraction is to reduce X-ray signals collected when the beam passes through a hole in the specimen (i.e., X-ray "hole-count").

The intensity of the signals from C and Ti rises at the interface as the electron beam enters the carbide particle (see Curves 1 and 4 in Figure 7).

(2) *Brackets.* Brackets may be used to enclose detail added by the writer to material being quoted. When they enclose commentary on what is being written, they often appear outside of the sentence.

The guidelines offer the following words of caution to new users: "Spherical and chromatic aberrations are present in every lab [transmission electron] microscope and will limit its performance." [detail added by the writer to a quoted statement]

The CO_2 seeping out of the flanks of Mammoth mountain has risen to nearly 1100 metric tons/day. [Mammoth mountain is actually a volcano.] [commentary]

24. Abbreviations and acronyms can save space and eliminate repetition. Many abbreviations, like the following, are standard usages: *K* (for *Kelvin*), *a.m.* or *A.M.* (for *morning*), *sq. ft.* (for *square feet*), *e.g.* (for *for example*), *Ph.D.* (for *Doctor of Philosophy*), and so on. You may look up most of these in a dictionary. Some abbreviations are formed by cutting a word off at a convenient syllable and adding a period. This kind of abbreviation is widely used as a way of saving space in graphics and bibliographies: for example, *config.* for *configuration* and *immunol.* for *immunology*. Don't use these in formal writing.

Acronyms, usually formed from the initial letters of a series of words, act as words in the sentence. Unlike abbreviations, they do not use periods. Some acronyms are standard usages and may appear in a dictionary, but many are invented by authors as convenient ways of cutting down on the repetitious use of lengthy terms. Generally, capitalize each acronym letter and do not use periods:

AAAS [American Academy of Arts and Sciences]
LDL cholesterol [low-density lipoprotein]
EDTA [ethylenediamenetetraacetic acid]
EMI [electro-magnetic interference]
POM [polarized optical microscope]
URL [uniform resource locator]

Enclose the acronym in parentheses after the words it stands for before using it by itself:

> Natural killer (NK) cells are controlled by receptors specific for polymorphic determinants of Class 1 molecules of the major histocompatibility complex (MHC). The origins of NK cells . . .

25. Use figures rather than words for numbers in scientific and technical prose. Always spell a number out if it begins the sentence. For very large numbers, use power expressions (e.g., 10^5). When numbers are modified by other numbers, spell out one number (e.g., *14 three-second pulses*). In nontechnical prose, spell out numbers for amounts of up to two words (e.g., *one hundred, nine hundred, four thousand*).

26. Italicize the titles of particular vessels, longer publications, words featured as words, and foreign words. Indicate italics in manuscript by underlining.

the Titanic, the Enterprise [titles of vessels]
Topics in Stereochemistry [book title]
the IEEE Transactions on Nuclear Science [journal title]
... the term spectral image refers to a ... [word featured as a word]
... is performed by a gene called sex-lethal, which acts ... [name of a gene]
... studies of L. donovani mutants suggest ... [genus-species name]

27. Use quotation marks to set off quotations and the titles of some short works.

(1) *Quotations.* When quoting directly from a source, use quotation marks. If the quotation is a complete sentence, capitalize the first word. Note that the second quotation mark comes after the period and before the footnote.

Quoting a complete sentence

As Smith predicted ten years ago, "All roads for membrane traffic may yet prove to be paved with the same kinds of molecules."[2]

Preceding a semicolon or colon

Jones states that the sludge is a by-product of "flue-gas desulfurization"; yet, we have traced it to the products of incomplete combustion. [The quotation mark precedes a semicolon or a colon.]

If you are quoting more than 50 words, use the indented block form to set off the quotation from the main body of prose. Block quotations do not use quotation marks.

(2) *Short works.* Articles, chapters, and sections from longer works, and titles of reports are commonly set off with quotation marks. Be sure to consult any relevant style guides for local practices.

Referring to a chapter from a work

Chapter 5, "Low-Carbon Structural Steels," treats steels with typical carbon contents in the range of 0.05 to 0.2%.

References

Adams, J. 1974. *Conceptual Blockbusting: A Guide To Better Ideas.* San Francisco: W. H. Freeman.

Adewusi, V. A. 1991. Enhanced Recovery of Bitumen by Steam with Chemical Additives. *Energy Sources* 13: 121–135.

Aviation Week and Space Technology. 1984. February 13: 75.

Cuadra/Elsevier. 1989. *Online Database Selection: A User's Guide to the Directory of Online Databases.* New York.

Department of Energy. 1991. Safety Evaluation Report: Restart of the K-Reactor, Savannah River Site. Supplement 3. DOE/DP-0093T.

Edelson, R. E., et al. 1979. Voyager Telecommunications: The Broadcast from Jupiter. *Science* 204(4396): 913–921.

Engineering Index, Inc. 1992. *Engineering Index Annual.* New York.

Engineering Information, Inc. 1992. *Engineering Index Thesaurus.* Hoboken, NJ.

Fenn, A. J., and G. A. King. 1992. Adaptive Nulling in the Hyperthermia Treatment of Cancer. *Lincoln Lab Journal* 5: 223–240.

Gelbart, W. M. 1982. Molecular Theory of Nematic Liquid Crystals. *Journal of Physical Chemistry* 86(22): 4298–4307.

Gunning, R., and R. A. Kallan. 1994. *How to Take the Fog Out of Business Writing.* Chicago: The Dartnell Corp.

Hardt, D. E., et al. 1982. Closed Loop Shape Control of a Roll-Bending Process. *ASME Journal of Dynamic Systems, Measurement, and Control* 104(December): 318--324.

Hurt, C. D. 1998. *Information Sources in Science and Technology.* 3rd ed. Englewood, Colo.: Libraries Unlimited.

Jagota, A., and P. R. Dawson. 1987. The Influence of Lateral Wall Vibrations and the Ultrasonic Welding of Thin-Walled Parts. *Transactions of the ASME, Series B: Journal of Engineering for Industry* 109(May): 140–146.

Kwack, E. Y., et al. 1992. Morphology of Globules and Cenospheres in Heavy Fuel Oil Burner Experiments. *Transactions of the ASME: Journal of Engineering for Gas Turbines and Power* 114(April): 338–349.

Merritt, M. W., et al. 1989. Wind-Shear Detection with Pencil-Beam Radars. *The Lincoln Laboratory Journal* 2(3): 483–52S.

Okerson, A., ed. 1991. *Directory of Electronic Journals, Newsletters, and Academic Discussion Lists.* Washington, DC: Association of Research Libraries.

Pillmoor, J. B., K. Wright, and A. S. Terry. 1993. Natural Products as a Source of Agrochemicals and Leads for Chemical Synthesis. *Pesticide Science* 32: 131–140.

Tufte, E. R. 1983. *The Visual Display of Quantitative Information.* Cheshire, CT: Graphics Press.

Voight, J. R. 1994. Morphological Variation in Shallow-Water Octopuses (Mollusca: Cephalopoda). *Journal of Zoology* 232: 491–504.

Whitney, D. E. 1982. Applying Stochastic Control Theory of Robot Sensing, Teaching, and Long-Term Control. Charles Stark Draper Laboratory. Report No. P-1314.

Index